A Non-Philosophical Theory of Nature

Radical Theologies

Radical Theologies is a call for transformational theologies that break out of traditional locations and approaches. The rhizomic ethos of radical theologies enable the series to engage with an ever-expanding radical expression and critique of theologies that have entered or seek to enter the public sphere, arising from the continued turn to religion and especially radical theology in politics, social sciences, philosophy, theory, cultural, and literary studies. The post-theistic theology both driving and arising from these intersections is the focus of this series.

Series Editors

Mike Grimshaw is an associate professor of Sociology at Canterbury University in New Zealand.

Michael Zbaraschuk is a lecturer at the University of Washington, Tacoma and a visiting assistant professor at Pacific Lutheran University.

Joshua Ramey is a visiting assistant professor at Haverford College.

Religion, Politics, and the Earth: The New Materialism
   By Clayton Crockett and Jeffrey W. Robbins

The Apocalyptic Trinity
   By Thomas J. J. Altizer

Foucault/Paul: Subjects of Power
   By Sophie Fuggle

A Non-Philosophical Theory of Nature: Ecologies of Thought
   By Anthony Paul Smith

A Non-Philosophical
Theory of Nature

Ecologies of Thought

Anthony Paul Smith

May you find some ideas of use.

palgrave
macmillan

A NON-PHILOSOPHICAL THEORY OF NATURE
Copyright © Anthony Paul Smith, 2013.

Softcover reprint of the hardcover 1st edition 2013 978-1-137-33587-6

All rights reserved.

First published in 2013 by
PALGRAVE MACMILLAN®
in the United States—a division of St. Martin's Press LLC,
175 Fifth Avenue, New York, NY 10010.

Where this book is distributed in the UK, Europe and the rest of the world, this is by Palgrave Macmillan, a division of Macmillan Publishers Limited, registered in England, company number 785998, of Houndmills, Basingstoke, Hampshire RG21 6XS.

Palgrave Macmillan is the global academic imprint of the above companies and has companies and representatives throughout the world.

Palgrave® and Macmillan® are registered trademarks in the United States, the United Kingdom, Europe and other countries.

ISBN 978-1-349-46328-2   ISBN 978-1-137-33197-7 (eBook)
DOI 10.1057/9781137331977

Library of Congress Cataloging-in-Publication Data

Smith, Anthony Paul, 1982–
    A non-philosophical theory of nature : ecologies of thought / Anthony Paul Smith.
        pages cm.—(Radical theologies)

    1. Ecology. 2. Nature. 3. hilosophy. I. Title.

QH541.S57 2013
577—dc23   2013002097

A catalogue record of the book is available from the British Library.

Design by Newgen Imaging Systems (P) Ltd., Chennai, India.

First edition: July 2013

10 9 8 7 6 5 4 3 2 1

Contents

Series Preface	vii
Acknowledgments	ix

Introduction		1

Part I The Perversity of Nature Foreclosed to Thought

1	Nature Is Not Hidden but Perverse	13
2	Ecology and Thought	19
3	Philosophy and Ecology	27
4	Theology and Ecology	45

Part II The Non-Philosophical Matrix

5	Theory of the Philosophical Decision	59
6	The Practice and Principles of Non-Philosophy	73
7	Non-Theological Supplement	95

Part III Immanental Ecology and Ecologies (of) Thought

8	Real Ecosystems (of) Thought	113
9	Elements of an Immanental Ecology	125
10	Ecologies without Nature	157

Part IV A Theory of Nature

11	Separating Nature from the World	167
12	Materials for a Theory of Nature	189

Conclusion: Theory of Nature	217
Notes	227
Bibliography	261
Index	271

Series Preface

Radical Theologies encompasses the intersections of constructive theology, secular theology, death-of-god theologies, political theologies, continental thought, and contemporary culture.

For too long, Radical theology has been wandering in the wilderness, while other forms of theological discourse have been pontificating to increasingly smaller audiences. However, there has been a cross-disciplinary rediscovery and turn to radical theologies as locations from which to engage with the multiplicities of twenty-first-century society, wherein the radical voice is also increasingly a theologically engaged voice with the recovery and rediscovery of radical theology as that which speaks the critique of "truth to power."

Radical Theologies reintroduces radical theological discourse into the public eye, debate, and discussion by covering the engagement of radical theology with culture, society, literature, politics, philosophy, and the discipline of religion.

Providing an outlet for those writing and thinking at the intersections of these areas with radical theology, *Radical Theologies* expresses an interdisciplinary engagement and approach that was being undertaken without a current series to situate itself within. This series, the first dedicated to radical theology, is also dedicated to redefining the very terms of theology as a concept and practice.

Just as Rhizomic thought engages with multiplicities and counters dualistic and prescriptive approaches, this series offers a timely outlet for an expanding field of "breakout" radical theologies that seek to redefine the very terms of theology. This includes work on and about the so-labeled death-of-god theologies and theologians who emerged in the 1960s and those who follow in their wake. Other radical theologies emerge from what can be termed "underground theologies" and also "a/theological foundations." All share the aim and expression of breaking out of walls previously ideologically invisible.

Acknowledgments

When I first began looking at scientific ecology the horror present in simplistic statements such as "everything is connected" became apparent to me. Saying that everything is connected is meant to convey a Green consciousness, but in reality the individual organisms present in those connections and the amount of material and energy they exchange is overwhelmingly complex. It can all seem too much!

It is with that sublime immensity of relationality in mind that I broach with fear and trembling acknowledging those who have assisted and supported me during the all phases of work on this book, completed during the course of my time at the University of Nottingham in the United Kingdom from 2006 to 2011. First, I must thank Hayley Smith for all the material support she has provided over the years without fully understanding why I care so much about these strange ideas and notions. My mother Laura Root deserves thanks for encouraging my weirdness early on, as well as my budding interest in environmentalism and writing when at the age of eight she helped me create a newsletter for my first grade class on Earth Day. Then there are my friends Bradley A. Johnson and Daniel Colucciello Barber who graciously read and commented on the draft chapters and papers that made up the original thesis. The ideas of both have marked me and encouraged me. Of course, strangely minor events, such as a drunken conversation or overcaffeinated debate, often play a major role in the creation of ideas. Many of these conversations are required for a book like this one, where many different fields of knowledge are engaged with and held together. For those I must thank friends and colleagues who settled in Nottingham for a time: Michael O'Neill Burns (for first arguing with me about Badiou and then later about Hegel), Thomas Lynch (for discussions about liberation theology and what it means to do theology and philosophy from a liberationist stance), Grant Whalquist (for discussions about perversion in Bataille and for encouraging me to think about how artists might receive these ideas), Sarah Fok (for discussions about creativity and applying philosophical concepts to problems outside philosophy), Catriona Gold (for discussions about animals and letting me know when the theory-jargon got too much for a critical geographer), Stuart Jesson (for discussions about nonviolence and generally being kind enough to listen), Alex Andrews (for discussions about general systems thinking and having done with a certain kind of theology), Karen Kilby (for always having a sympathetic ear and for the trust and guidance she extended to me regarding teaching), and Orion Edgar (for discussions, usually whisky fueled, about the place of religious beliefs and practices and what they mean to him in relation to nature, ecology, and food).

Of course dear friends outside Nottingham also played a role as sounding boards for ideas. Nicola Rubczak, whose brilliant work encouraged me to consider feminist ways of thinking, was an immense help in dealing with issues in the French. Daniel Whistler, whose philosophical acumen I greatly admire, has been a tremendous help with short but brilliant "tutorials" (often given on buses or trains or, on one occasion, short walks around the strange but beautiful English countryside) on Spinoza, Fichte, Schelling, and the relationship between philosophy and religion. Marjorie Gracieuse for all the discussions about Laruelle and Deleuze as well as calming me down when I was anxious about my viva. Liam Heneghan, the philosopher stuck in an ecologist's body and codirector of DePaul University's Institute for Nature and Culture, who not only gave me my first paid academic work when I was an undergraduate student in philosophy at DePaul University, but also became a close friend and helped me navigate the world of scientific ecology. Two other DePaul friends deserve to be thanked as well, though not for any direct contribution to the book: William Jordan III helpfully slowed me down early in my research by putting forth his own ponderous questions to me; and Peter Steeves provided not only personal encouragement but also intellectual inspiration. Then there is Adam Kotsko, who I have to thank for creating our little online community of support, intellectually and financially, at the blog *An und für sich.*

I was encouraged to turn the original thesis into a book, first, by my PhD examiners Dr. Steven Shakespeare and Prof. Agata Bielik-Robson, whose tough examination humbled and emboldened me, as does our continued friendship. However, it was the prodding and counsel of Joshua Ramey that finally allowed this book to see the light of day. I am immensely happy that he, along with his coeditors Mike Grimshaw and Michael Zbaraschuk, wanted to include this book in their *Radical Theologies* series. Burke Gerstenschlager at Palgrave has been very supportive and made the publishing process feel human, a rare achievement in this business and testament to his skills as an editor.

Finally, I continue to be amazed at the openness and kindness François Laruelle has extended to me, always patient with my broken French and

ACKNOWLEDGMENTS   xi

persistent questions. Reading Laruelle opened up thinking for me and made it exciting once more. This experience was a repetition of an earlier event in my life, one that has marked the trajectory of my life for the past 8 years. I'm thinking of the experience I had at the age of 20 when I was living in Paris and reading—foolishly doing so in English!—Philip Goodchild's *Capitalism and Religion: The Price of Piety*. I knew then that I wanted to be his student and I thank him for taking me on as well as for all the guidance, freedom, and inspiration he has given me.

This list includes just humans! There is really too much to acknowledge, too much to affirm, even in the heartache I experienced in my five years at Nottingham surrounded by the jackals of radical orthodox theology. This is the horror of everything being connected, but also a kind of horror or shame at the amount of love that I have experienced in the midst of that heartache and frustration. But I know this work is a product of the expression of this love and grace, which is an exchange of love and grace between all these friends and me. For this love and this grace are hidden in the apparent words and thoughts that form this thesis. This is radical immanence.

Earlier versions of some of the material that makes up parts I and II have appeared in sections of "Thinking from the One: Non-Philosophy and the Ancient Philosophical Figure of the One," in *Laruelle and Non-Philosophy*, edited by John Mullarkey and Anthony Paul Smith (Edinburgh: Edinburgh University Press, 2012), pp. 19–41; "The Real and Nature: A Heretical Nature contra Philosophy's Nature," in *Identities: Journal for Politics, Gender, and Culture* 8.2 (2012), pp. 55–67; "A Stumbling Block to the Jews and Folly to the Greeks: Non-Philosophy and Philosophy's Absolutes," in *Analecta Hermeneutica* 3 (2011), pp. 1–16; "What Can Be Done with Religion?: Non-Philosophy and the Future of Philosophy of Religion," in *After the Postsecular and the Postmodern: New Essays in Continental Philosophy of Religion*, edited by Anthony Paul Smith and Daniel Whistler (Newcastle-upon-Tyne: Cambridge Scholars Press, 2010), pp. 280–298; and "Philosophy and Ecosystem: Towards a Transcendental Ecology," in *Polygraph* 22 (2010), pp. 65–82. I gratefully acknowledge the permissions granted to use all this material.

Introduction

There is a very old philosophical story. We no longer have the full tale, but only the ending that goes, "Nature loves to hide." We don't know if this is a drama or a tragedy. If you say it with the right inflection it could even be the punch line to a joke. But is the joke philosophical or is it a joke on philosophy?

There is another story, even older, and this time theological. It begins with the beginning (though whether or not it is the very beginning is up for debate) saying, "In the beginning God created the heavens and the earth." The story goes on to say that at that time the earth was formless and would go on being nothing without this God. Now, they say this story has an ending and we see there that the earth returns, "Then I saw a new heaven and a new earth; for the first earth had passed away, and the sea was no more."

This story may be epic in scope, but we still don't know if it is a joke or not. And we don't know who that joke is on. For nature has become a problem. Small talk is filled with discussion of this problem, for now when we talk about nature we know that we're talking in part about global climate change. Some may say that it is shameful how human beings are unable to work in harmony with nature. That we go against the order of nature. But this present work began with a different thought. It began with the thought that nature has become a problem for nature. For if human beings are natural then there is nothing unnatural about what we are doing to the biosphere. We haven't risen above the natural in the creation of nuclear power any more than the beaver does when he constructs his dam. The problems caused to other living things by human pollution are entirely natural. For what is, is natural. There is simply no other test for it. It's not as if we can name something an aberration of nature by measuring how long it can persist living. Duration of living is relative after all and we know that those creatures living now are but a tiny percentage of those who have lived on this planet. Yet they were natural and their extinction too was natural. It is natural to be born, to persist, and to die and it is natural to the acts that occur during that persistence and even the "corruption," what we now call decomposition, that follows after death.

Nature is also a problem at the level of ideas. For the idea of nature seems to have no direct object. I am unable to simply put nature in front of me anymore than I can put Being or love in front of me. Rather than attempting to think nature under this impossibility of objectification, we instead see in the history of thought a proliferation of very different ideas of nature. There is the Romantic understanding of nature, where one is truly free and all is in balance, mingling openly with the pessimistic understanding of nature as the site of suffering, pain, and death. There is also the theological understanding of nature as a good creation, which stands side by side with the notion that nature is organized hierarchically such that domination and even death is taken to be not just natural but also divine.

What interests me is not entering into philosophical and theological debates about what nature could be. These seem to me to be intractable debates that are more concerned with the philosophical or theological project putting them forward rather than with thinking nature outside of those projects. No, what interests me is rather that these projects, the proliferation of strikingly different ideas concerning nature, *are natural*. And so in this work I attempt to think and develop that very idea. Subsuming thought into nature, but without becoming a naturalism or materialism, without reducing ideas to some kind of "mere nature," for this itself is but one of those ideas about nature.

The method for this work is derived from François Laruelle's non-philosophy as he has developed it over the last three decades. This method and the way I've synthesized it is explained at length in part II. What Laruelle's non-philosophy allows me to do is envisage a different organization of thinking within different practices of knowledge. Rather than the usual division of labor between a science that does not think and a philosophy or theology that must think for science and thereby make sense of its stumbling across reality, Laruelle fosters a democracy (of) thought that includes philosophy and science expressed also as a unified theory of philosophy and science. Even the way he writes this democracy (of) thought, with the suspended "of," speaks to this practice of a real democracy as the relationship between democracy and thought is itself suspended and becomes instead as if One. And the figure of the One is also of great importance here, as our attempt to think a nature that brings about the effects of the ideas of nature is also an attempt to think nature prior to Being and Alterity. To think what Laruelle calls the radical identity of nature as if One.

However, there is also a model for the work that I have used but that I don't discuss at any length. I suspect those who know it can feel

INTRODUCTION   3

its influence, but it remains somewhat concealed. That model is Philip
Goodchild's work as developed in his *Capitalism and Religion: The Price of
Piety* (2002) and *Theology of Money* (2009). The second book develops a
theory of money using the tools of philosophical and theological analysis
in dialogue with economics, in much the same way that I develop my own
unified theory of philosophical theology and ecology to construct a theory
of nature. In the first book Goodchild, who was of course my doctoral
advisor, creates a critical theory of piety by using a multitude of thinkers
as catalysts for thinking. The work of each thinker is not presented in a
purely scholarly fashion, rather each thinker is presented in that book as
a condensed crystal. These "readings" are not reductive, however, but are
intensive and so it would be a mistake to read that book with the intention of learning how to do historical scholarship or historical readings on
Deleuze or Marx or Spinoza or whomever. Rather, though this is mixed
with the language of Laruelle, each thinker becomes a simple material that
is put into the process of thought reorganized around attention to that
which matters most. What matters most according to Goodchild is suffering. Suffering also directs my attention here, though it may be obscured
somewhat through the form it has taken.

For, as an academic work, it tends toward a certain distance from exis-
tential concerns raised by what is commonly called the environmental crisis. I myself prefer not to refer to this ongoing event as a crisis, since the
word "crisis" suggests that there is some point in time and space where
certain forces will converge, or more realistically *have* converged, and thus
we may be led to think, in some heroic fantasy, that human beings can
marshal a kind of army against the forces of destruction and turn back this
crisis. The reality is more distressing. For the so-called environmental crisis
is now more our everyday reality as climate, as even newspapers move from
discussing natural disasters to the normalization of "weird weather." As
such it feels too ordinary in its weirdness to be a crisis. Moreover, the force
of destruction is simply the political and social organization of human life.
This is an organization that in many ways *works* and because of this productive power as organization we see no way out. It also may seem like this
work is distant to suffering because of the high level of abstraction going
on. The project is ostensibly concerned with the most abstract, "thought,"
and it goes about thinking with an incredibly abstract system that is in
many ways alien to standard philosophical and theological forms of thinking. Yet, this alien nature is intentional. It is an attempt, begun early in
my studies, to think as immanently as possible and to find in that way of
thinking not a thought without an exterior, which would set up any end
to suffering as an impossibility, but to think the immanence of nature as
both real and the source and outworking of a resistance to suffering and

4   A Non-Philosophical Theory of Nature

the powers of this World. This only seems strange, I claim, because what is most immanent to us is also most alien precisely because of the hallucinatory transcendent structures of thought developed in philosophy and theology. And to break from those transcendent structures I infect them with material from scientific ecology. This mutation causing infection should not be confused with raising science to the level of the sole arbiter of truth. It would be counterproductive to simply reverse the hierarchy of science and philosophy/theology as if science was to become some kind of priest for immanent divinity. But science, I argue following Laruelle, has a particular posture in its practice, one that is paralleled in theology to some extent, that thinks from its object and changes its practice on the basis of its relation to that object. This posture is an expression of radical immanence.

So what is this immanence? What does radical immanence mean in this work? The focus on immanence in recent Continental theory is due in part to the importance given it in the work of Gilles Deleuze, but the term was central for the phenomenology of Edmund Husserl as well. A focus on immanence can be found at the heart of widely divergent thinkers such as Alain Badiou, who understands his philosophy to be materialist, and immanence can be found in virulent antiscience theological philosophies such as that of Michel Henry's. In each thinker the meaning of the word is different, but in its most common Deleuzian form it is sometimes confused by some English-language commentators as being a variant of naturalism. This interpretation is shared in a hostile register by Radical Orthodoxy theologians. Thus they take the notion of there being no absolute exterior to immanence to mean that there can be no appeal to something other: there is nothing but what there is.

In my view this reading of Deleuze is wrong, but dealing with those readings and presenting a counter-reading is outside the scope of this project, which begins instead with Laruelle's understanding of immanence. He himself differentiates this from what he takes to be Deleuze's understanding for Deleuze, according to Laruelle, attempts to think an absolute immanence whereas Laruelle thinks from a radical immanence. Radical immanence is the experience or style of thinking in-One or from the Real. Laruelle explains his theory of immanence, differentiating it from the philosophical "misadventures" of immanence writing, "In philosophy, Marxism included, immanence is an objective, proclamation, an object; never a manner or style of thinking."[1] Radical immanence is different from the Nature of naturalism; it is different from a quasi-thing above consciousness or humanity. Thinking in the manner of radical immanence is to think neither as a part in a whole, nor as a cog in a cosmic machine, but in a manner already-manifest prior to thinking as inscribed within a

INTRODUCTION   5

system. For that system itself is produced from that immanence. Radical immanence is prior to ontology or to the difference or alterity to ontology. Thus we can speak of the immanence of what is, but this "what is" is never a "merely what is." For radical immanence is what is, but it is also what could be and what could have been but will never be. More than this, it is lived. It is the thought thinking more than it is the totality of thought. It is not a system, but neither is it unknowable. It is both fleshly, because a creatural body is radical immanence, but it is also the potential for that body to die and to love. It is the unconcealed prior to knowledge, prior to theory, and theory can never circumscribe it but can recognize that it already is it. This conception of immanence avoids a kind of ideational friend/enemy distinction between itself and transcendence. This is true for two reasons: first, transcendence is understood as an effect or a production of immanence from this position because, second, transcendence is in some sense Real at the relative level. There is no reason from the perspective of the radical immanence of a lived body that what is must be. There is a relative transcendence at work in the life of that lived body, but that relative transcendence must be thought in the manner of a lived body. Nature as a transcendent, quasi-thing is thought in the manner of radical immanence and in this way nature is thought in exile from worldly Nature, a wandering Stranger, rather than one who is at home and content with the mere what is.

This brings us then to the structure of the book. I have split the work into four parts. In part I, I investigate the standard relationship between scientific ecology and both philosophy and theology. I begin there with the axiom: nature is perverse. This is to say it outruns thought. As a result, each regional knowledge or knowing (the gerund is intentional) is in an equivalent position with respect to nature: each remains partial and incomplete. Beginning with this axiom already separates the work undertaken here from classical Greek metaphysics, the transformation of that metaphysics in Christian theology, as well as the contemporary forms of reductionist naturalism and materialism. In this section I show how ecology's object is not nature as such, but the ecosystem. The ecosystem reveals aspects or occasions of nature, but is not nature itself, for nature outruns ecological thinking as well as philosophical and theological thinking. I then turn to the respective relations between philosophy, theology, and ecology in order to show the various types these dominant standard relational forms may take. Philosophy has a subsumption type (which includes, at least on most readings, Plato to Hegel, Husserl, and Heidegger), where science must be subsumed into philosophy in order to think, and a bonded type (which includes Aristotle to Schelling and Žižek), where philosophy claims to be bonded with science in its thinking. Theology is located within the subsumption model, but finds that it can either be a declension type (which includes, interestingly, Karl Barth and Pope Benedict XVI), where science can only point to its fallenness, or an inflection type (which includes most ecofeminists and the liberation theology of Leonardo Boff), where it can transform itself by working with what it sees as the best, most theologically fruitful science. In each case, though, there is a certain shallowness in terms of engagement with scientific ecology. For many philosophies this is often in the name of some kind of naturalism or materialism taken from a bonded model with physics. Yet many of the subsumption philosophies and theologies also purposefully avoid engagement with ecology, as with all science, as it aims to avoid the perceived failings of both naturalism and materialism. Even when an inflection type theology thinks with science it does so with the aim of transcending the shortcomings of naturalism and materialism. The theory developed in the final part of the book also aims to avoid naturalism and materialism. Or, rather, in precise terms, I aim for a theory of nature that is not overdetermined by naturalism or materialism but one that can nevertheless use naturalist and materialist philosophies in its construction of a theory of nature just as it can use those philosophies and theologies antagonistic to naturalism and materialism in the same way. The survey and typology provided in part I thus serves two purposes: first, to show the failure of philosophy and theology to engage in a deep way with the material of scientific ecology and, second, to locate the ideational fields (philosophy, theology, and ecology) that will be our sources in parts III and IV for the construction of a non-philosophical theory of nature.

After this survey and typology of the dominant standard relational forms, I turn in part II to a deep explication of François Laruelle's non-philosophy. Laruelle's work is renowned for its difficulty, yet that difficulty is but an expression of an attempt to think what is very difficult—the radical immanence of the One without recourse to subsuming it into Being or Alterity. While Laruelle's work has been lumped in with a general trend in Continental philosophy that sees a return to engagement with science, sometimes given the misnomer of "Speculative Realism," his project differs markedly from others grouped under this name, like the science and technology studies of thinkers such as Bruno Latour as well as the recent "return to the Real" in thinkers such as Quentin Meillassoux. The term "non-philosophy" often gives the reader a sense that this is a philosophy that says no. Really though, the "non" in non-philosophy takes its cue from the "non" in non-Euclidean geometry, and like non-Euclidean geometry it does not negate philosophy, but thinks philosophically in a different, more general way according to different axioms. Laruelle's work provides a model for how the theoretical humanities can come together with scientific thinking while avoiding the pitfalls of positivism (which

INTRODUCTION 7

has certain structural resemblances with naturalism and materialism) but still remain a transcendental practice (which he will go on to call an "immanental practice"). As Laruelle's work is relatively unknown in the English-speaking world and because it provides the methodology for the project, I have devoted a large section of the work to explaining the practice of non-philosophy. I begin with his theory of the philosophical decision, which at first appears as a simple criticism of philosophy but actually functions in non-philosophy as a location of philosophy's unacknowledged limits and its identity. Locating these limits is what allows Laruelle to construct a different way of thinking alongside scientific material while doing something akin to traditional metaphysics. This metaphysics is divorced from the usual absolute focus on Being and without recourse to an absolute transcendent Alterity to break with the obsession with Being. Following this explication of the philosophical decision I turn to the form this metaphysics takes, which Laruelle calls a "philo-fiction." Of real importance in this chapter is the way this philo-fiction organizes a way of thinking that does not aim to be absolute and allows for a mutated form of philosophical thinking by way of scientific material. I then end part II arguing for what I call a non-theology that deals with a potential weakness in non-philosophy's practice, insofar as it seems to mirror the theological structure of thinking in a secular way. Non-Theology complements non-philosophy by allowing it to engage with and mutate theological forms of thinking more intentionally than we find in Laruelle himself.

I then put the method of non-philosophy to use in part III, specifically with regard to what Laruelle calls "unified theories" of philosophy and some material outside philosophy. In this instance it is a unified theory of philosophical theology and ecology or the construction of what I call an immanental ecology. Here I follow Laruelle in his latest work where he engages with a specific scientific material, which for him is quantum physics, within the wider philo-fiction of non-philosophy. Part three is bookended by discussions of ecology and philosophy, specifically with regard to Laruelle and then with ecological thinkers Latour and Timothy Morton who want to be finally rid of the idea of nature altogether. But between these bookends I give special attention to the technical aspects of scientific ecology, centered on the concept of the ecosystem, and try to think through their consequences for a philosophical theology. While the principles and practice of non-philosophy are what allows us to do this with a certain rigor and consistency, the goal of this section is to provide the conceptual tools necessary to think of thinking itself as ecological. Ideas about nature found in philosophy and theology can now be treated as if they could be explored ecologically, rather than ecology requiring a philosophy or theology as such. The aspects or occasions of nature disclosed

8   A NON-PHILOSOPHICAL THEORY OF NATURE

in ecological principles and concepts, derived from the study of the ecosystem, challenge some standard philosophical and theological notions of
nature. However, instead of focusing on this challenge, which in many
ways would repeat some of the criticisms already made in part I, I focus on
trying to think philosophically and theologically from within the scientific
material. So, this challenge is presented not to shut down philosophical
and theological thought, but to actually be productive of a different way
of thinking within a unified theory of philosophical theology and ecology
or immanental ecology.

This comes to fruition in part IV. Immanental ecology allows me to
treat philosophical and theological conceptions of nature as simple material that can be reorganized into a new, more adequate theory of nature.
This comes out of a recognition of the perversity of nature as well as the
natural processes of the ecology (of) thought explored in earlier chapters.
It is important to understand that I am not claiming here that ecology
provides a more adequate understanding of nature and that philosophy
and theology need to simply explicate what ecology already thinks implicitly. This would again be a standard form of philosophy or theology that
understands the scientific material to be in some way separate from the
philosophical or theological practice. My claim is rather that nature, as
perverse, can never be thought fully by any single discourse and in fact
can never be thought fully. Instead I claim that the theory of nature constructed here is adequate to that reality, that a unified philosophical theory
and ecology can think nature otherwise than we find in naturalism. To
make that argument I first show that the various proliferating ideas about
nature are all thought under an unthought dominance of the World in
philosophy and theology. This worldliness of thinking about nature is
located as an invariant in philosophy that goes mostly unacknowledged. I
then turn to a philosopher who has made an investigation of World central
in his work, Martin Heidegger, and his use of the fourfold to think worldliness. The fourfold, perhaps one of Heidegger's more daunting concepts,
is treated under Laruelle's non-philosophy (specifically a practice called
"unilateralization," which is explained in part II) to create a single dyad
made up of two minor dyads. These dyads come to stand in for what is
commonly referred to as nature and culture, but as understood through a
relationship of veiling/unveiling and presence/absence. These relationships
are in many ways just as theological as they are philosophical, and indeed
there is a deep connection in the nature dyad with the way the relationship between God and nature has played out in theology and philosophy.
I then turn to these various forms of relation that these dyads exhibit in
theology and philosophy by looking at St. Thomas Aquinas, Benedict de
Spinoza, Abu Ya'qûb al-Sijistânî, and Naṣīr al-Dīn al-Ṭūsī. What is given

INTRODUCTION   9

in the presentation of these thinkers are not readings as such, but rather I treat them as processes that can be treated ecologically precisely because thought in general has already been shown to be ecological. I call these processes, following Goodchild, an expression of piety. As this is a unified theory of philosophical theology and ecology I am able to think of piety as something ecological and ecology as being traversed by piety. For in the unified theory both the terms of ecology and the terms of philosophical theology are mutated. In this way I am not entering into a wider scholarly debate, but instead I am treating these thoughts as ecosystems that have material and energy that can be extracted.

Finally, in the conclusion, I construct a non-philosophical and non-theological theory of nature using the materials and energies extracted. This theory can be summarized as having a tripartite structure that understands the creatural as subject of nature, the chimera of God or nature as non-thetic transcendence of nature, and the One as radical immanence of nature. This summary is in terms that will here, in the introduction to this work, seem impenetrable, but by the end of the book should come to have a determinate meaning for the reader. However, the general shape of this theory can be understood if you consider what each aspect of the theory avoids. For, by thinking nature as One or nature from radical immanence, I am able to avoid both the subsumption and bonded model of philosophical and theological thinking. For philosophy and theology subsume nature into their own ontological concerns and thus confuse nature with Being. This may be why philosophers have claimed that nature loves to hide, for they never actually looked for it, instead preferring to look for some matter (materialism) or normative idea (naturalism) that nature would be reducible to. By thinking the chimerical identity of God or nature as a non-thetic transcendence of nature I am able to avoid naturalism more fully, or rather I am able to show that the transcendent, hypostasized conception of nature operative in naturalism is actually a production of immanence and thus is itself ecologically produced rather than providing metaphysical rules for ecological processes. In this way a space for freedom is opened up. This space is not opened up "in nature," for freedom is natural, but is rather opened up in the sense that the transcendent aspect of nature is ultimately made relative to a lived immanence common to all creatures. This creatural aspect of nature is what I claim is the subjectivity of nature. Nature too has a subjectivity produced from its radical immanence such that every creature is said to be natural. This disempowers any strong sense of normativity in "the natural" such as we find in Thomistic natural law theory. For the perversity of nature is present as subject in the ongoing and diverse creation of niches by species (which are immanently connected, as you will see in chapter 9). Indeed there is a certain messianity present in this subjective character of creatures, for as creatures constantly proliferate and spring forth into the biosphere they reject the predictability of naturalism's *sékommça* or "that's how it is." In a very real sense the plethora of these creatures, as the subject of nature, speaks to the destruction of every "law of nature" (really an undue anthropomorphism) in the name of a single law: that of the sabbath or a radical, if not absolute, freedom.

In short, what is created in this theory of nature, derived from the construction of a unified theory of philosophical theology and ecology, is a nature that cries out violence against worldly Nature. For worldly Nature conceived in a quasi-theological manner by naturalism is but a hallucination of an absolute. It presents itself as the cold measure of everything, under which we would have no right, no power, or potential to protest against it. But when this aspect of nature is shown to be a production of the radical immanence beyond Being and Alterity, then the creatural subjectivity of nature itself protests against the violence of this Nature. This nature is then a Stranger, having rejected servitude to worldly Nature it enters into exile from the World and wanders the earth without any transcendent roots. This exile of nature, an exile that frees us not only from bad and nearly forgotten jokes, is the common condition of every creature, from the human being awaiting the advent of the messiah to the leaf turning itself toward the sun to receive its energy for the day.

Part I

The Perversity of Nature Foreclosed to Thought

Chapter 1

Nature Is Not Hidden but Perverse

In the introduction I said that the goal of this work is to foster a democracy (of) thought among the disparate fields of philosophy, theology, and ecology. This democracy (of) thought is not an end unto itself, but is necessary in order to denude these discourses of any pretense to a hierarchical posture over the others. This in turn will allow us to treat material within these discourses as just that—simple material that can be distributed and organized in a different ecosystem (of) thought. This chapter serves to survey these fields as they are currently organized in relation to one another. In terms that will be discussed at length in part III, we will examine the ecotones or the limits of their identity as they come up against one another as already constituted, though unconsciously, as ecosystems (of) thought (an ecotone is a transition zone between two different ecosystems, often there will be a blending of elements from two different ecosystems and species will be present in the ecotone that are not present in either of the two bordering ecosystems). I will trace their limits and the spaces at their limits where they blend (ecotone) and in these limit-ecotone spaces we will find what remains unthought within their strict borders, what remains presented as if unecological in being thus thought, and we will then begin to identify the perversity of nature foreclosed to thought. As we will come to see, it is this blindness of these discourses to the perversity of nature foreclosed to thought, their refusal or inability to allow scientific ecology to infect and mutate their own thinking about their own thinking, that lies behind their remaining unecological in thinking nature.

Recognizing the perversity of nature is recognizing that nature is stranger than any one regional knowledge, be it philosophy, theology, or scientific ecology, can capture. Recognizing the perversity of nature means recognizing the radically foreclosed character of nature to thought. In terms that

14   A Non-Philosophical Theory of Nature

I adapt from Laruelle, "nature" becomes a first name for the Real. Laruelle gives this definition to first names, "Fundamental terms which symbolize the Real and its modes according to its radical immanence or its identity. They are deprived of their philosophical sense and become, via axiomatized abstraction, the terms—axioms and theorems—of non-philosophy."[1] A certain term, chosen in part for its fittingness with the Real, is transformed from its philosophical sense or meaning and thought according to certain axioms of the Real. The sense that nature is foreclosed to thought is a mutation of the historical philosophical stance toward nature, transmitted through its Greek filiation, that "nature loves to hide."[2] As Pierre Hadot has shown, this underlying idea about nature has been able to accommodate a variety of very different philosophical visions about nature, from its original meaning in Heraclitus that the death of things is unavoidable, "What is born tends to disappear," to the modern antagonism between the Promethean and Orphic attitudes.[3] The first, combining the attitudes of both magicians and scientists, claims that nature has hidden itself in mechanization and that mechanics itself can unveil nature and reveal its secrets, which are of or can be turned into human use. The second, sharing much in common with the green notion of "small footprints," seeks to unveil the spiritual secrets of nature, to unveil nature though contemplation, art, and poetry and thereby take pleasure in this knowledge without any particular concern for its use. While there is certainly an antagonism between these two attitudes, there is also a fundamental amphibology: for both, nature is veiled and can be unveiled.

This is not what our axiom, that the perversity of nature is foreclosed to thought, means. Nature itself is not veiled, nature does not "love to hide"; no, nature is radically immanent as the Real. That is, the metaphor of nature's veil already beguiles one into thinking that there is something other than nature, something we can appeal to outside of nature. Yet, if we think nature in an ecological thought we have to recognize that the veil is also nature! No, nature is not veiled, but thinking this allows our regional knowledges to think that they can unveil nature, that they can touch and circumscribe nature with thought and thereby either exploit her for our own gain or save her. Our contemporary climate, both in the physical and intellectual sense, is determined by a single force: the neoliberal capitalist ideology that demands everything reduce its value to the quantitative measure of money so that it can produce more of this measure. Nature, though, appears to be purposely deviating from what is accepted as good, proper, or reasonable in capitalist society. Nature itself appears to be refusing to go away, to separate itself off from "culture" and the human person, and insists on inhering to every part of culture and within every human person, and it resists bowing before capitalism's demand, to be measured as

NATURE IS NOT HIDDEN BUT PERVERSE   15

something relative rather than the radical condition for any relative measurement.[4] This is not hidden from us; we know the perversity of nature. It is present in our bones, the aches some get when a storm is coming and the way that weather is no longer a matter of mere conversation but of life and death concern.[5] We are witnesses to the perversity of nature as we are an instance of its perversity.

In other words, which will be explained in more depth in part II, the causal relationship of nature to thought is a unilateral one. Nature, as a name for the Real, determines all thought; in the last instance all thought is natural. This may cause certain misunderstandings. For instance, someone may read this and think it means that thought has no influence or causal power in the world. This would be to confuse two levels of autonomy, the relative and radical, for, of course, thought can affect things in the world. An idea can lead to or participate in a change to a society. An idea can lead us to destroy an ecosystem or to restore a degraded one. Yet, none of this destroys or saves nature as such. The thought can never become unnatural; it is never not a real idea and what is real is natural. Thought can have real effects, but cannot affect the Real; thought can think the unnatural, but it does not do so unnaturally.

Allow us to step back for a moment, before diving into the local material of specific thinkers, and survey the whole of the field from a little higher up. We have three distinct regional knowledges, what we call ecosystems (of) thought: philosophy, theology, and ecology. These identities may seem too pure in the simple separation here, for, as regards philosophy and theology, there has been no actual purity of either that we can locate in the history of thought and the same holds true for ecology, as it found itself developing among and responding to philosophical and theological notions of nature. The messy reality of these discourses gives me no offense and it does not need to lead into mystification. After all, though Spinoza devotes the first part of his *Ethics* to a treatise on God, surely a theology by definition, no one feels all that uncomfortable calling him a philosopher. In the same way Aquinas, while clearly devoting much of his work to "pure" philosophical matters or matters that seemed removed from the everyday problems of religious believers, he is nevertheless a Doctor of the Roman Catholic Church and we have no difficulty referring to him as a theologian. Finally, though Aldo Leopold's classic 1949 work *A Sand Country Almanac* bears upon certain philosophical problems, both metaphysical and ethical, he was never a professor of philosophy but rather was a forester and eventually became a Professor of Game Management in the University of Wisconsin-Madison's Department of Forest and Wildlife Ecology and no one seems to think a professorship in philosophy was stolen from him.

We are comfortable calling the work of one a "philosophy," the other "theology," and the third "ecology." There is a deeper reason for that comfort than mere institutional status, where they did their work or where that work is now taught, and it goes to the heart of their identities as distinct practices of knowledge. Their identity as philosophy, theology, or ecology has to do with the material of thought that they work with and the way in which they work with that material. For example, both Aquinas and Spinoza wrote about God and nature, in both cases their material can be said in the abstract to be the same and both even engaged significantly with Christian and Jewish scripture, but their stance toward that material and subsequent practice differs radically. Aquinas approaches problems from the perspective of a Christian theologian; all his work is ultimately concerned with the particular reception of revelation within the Roman Catholic tradition. Thus, when it comes to nature Aquinas himself recognizes that the philosopher and the theologian think in very different ways, the theologian according to the "light of doctrine" while the philosopher considers creatures (creation or nature) "as they are." Aquinas explains,

> The Christian faith, however, does not consider them as such; thus, it regards fire not as fire, but as representing the sublimity of God, and as being directed to Him in any way at all. For it said: "Full of the glory of the Lord is His work. Hath not the Lord made the saints to declare all His wonderful works?" (Eccles. 52:16–17)[6]

Spinoza, insofar as he is not working within a particular community of faith and aims at a universal knowledge that undercuts conflicts concerning the specifics of dogma, is quite different. While in the aforementioned passage we see Aquinas ground his distinction between philosophical and theological metaphysics in a passage from scripture, Spinoza's considerations of scripture as revelation lead him to posit the superiority of natural knowledge over revealed knowledge. He writes, "I prove that the revealed word of God is not a certain number of books but a pure conception of the divine mind which was revealed to the prophets, namely, to obey God with all one's mind by practicing justice and charity."[7] This is revealed knowledge in the sense that it is given to the people from positions of authority, but revealed knowledge does not clash with natural knowledge, which is equally divine.[8] In fact, for Spinoza there is nothing in revealed knowledge, as claimed by particular traditions that cannot be known more securely in natural or universally revealed knowledge.

In some sense, then, when we use theology in the course of this work, we refer to a relating of everything back to God as understood within a community that has arisen around a specific understanding of a revelation.[9]

NATURE IS NOT HIDDEN BUT PERVERSE   17

The material is reality itself, but the posture taken toward reality is determined by the development of dogma in the light of a particular revelation, while philosophy, especially at its limits in thinkers such as Spinoza, aims to think from a position that it takes to be more universal, unmoored by strict boundaries (though there are of course some) and to find some kind of secure grounding for knowledge outside of particular or local revelations. Here the material is also reality itself, but the stance is more universal, in varying degrees, and an attempt to think a universal ground of knowledge. This means that philosophy can dismiss more easily certain antagonisms between certain dogmatic statements coming out of a religious community and what knowledge derived from "nature" or from outside of that religious community, but it also means that philosophy tends to split up thought itself in a way that theology does not tend to, for instance, between revelation and ground. These are both incredibly schematic definitions and not intended to bestow any kind of absolute judgment on either theology or philosophy, but only to delineate distinct fields by way of strong tendencies in terms of material worked with and the practice of working on that material. As will become clear throughout this chapter I find within both fields aspects that are problematic in terms of thinking ecologically and aspects that are indispensable. We can, however, give a definition to ecology that is a bit more precise. While the material of philosophy and theology is reality itself, a necessarily abstract definition if we are to include all the various philosophers and theologians valued as such in the history of thought, the material for ecology is more concrete and common among ecologists. The primary material is that of the ecosystem, discussed at more length later, and it is from the concept of the ecosystem that working with any other material is practiced, be it philosophical or some physical material within a particular environment.

These then are our three distinct regional knowledges that we move within in this work. We can speak of their limits as regards each other because of the dominant tendencies we have located, which also avoid any kind of naive, strict separation or desire for purity among them. What is most at issue is not the relationship of philosophy and theology, often an antagonistic one that every philosopher and theologian has some opinion on. No, what is most at issue here is the relationship between science and philosophy or theology, specifically between ecology and philosophy or theology. Not as regards the historical relationship between science and philosophy or theology, which has been both antagonistic and beneficial, but as regards this specific science, ecology, and the stances that philosophy and theology take toward it with regard to their own thought. First, we will examine the relationship of ecology to philosophy or theology, which I simply call "thought" in this section. This is important because the overdetermination of scientific ecology by prior philosophical or theological images of thought tends to go unacknowledged. Corollary to this, many miss the fact that certain key concepts in ecology are responses to that overdetermination and therefore these concepts could be taken up within philosophy and theology as well. I will then turn to the relationship of philosophy and ecology, and a short section on recent philosophies of nature's "bondedness" with physics, while saving a wider discussion of philosophy's relationship toward science in general until the next chapter.

While I discuss a variety of different philosophical positions, from those whose work is self-described environmental philosophy to phenomenology and new philosophies of nature, they can be separated into two general types. The first I call the subsumption type, where science must be subsumed into philosophy for it to think, and the second I call the bonded type, where philosophy is understood to be bonded to natural science, specifically physics, in its own philosophical operations. With regard to the second I will show that philosophies of nature have not engaged with ecology, limiting their scientific contamination and in the first I will examine the very limited engagement with ecology that environmental philosophy has. Finally, I will turn to theology and ecology and trace similar limits, though understanding the relationship of science and theology requires a different typology. While some environmental theologies have some sense of a "bonded" element with ecology, this is always relative to a theological subsumption that is inherent to theological practice since all things must be related ultimately back to the divine. Thus I differentiate two types within this theological subsumption of science: the declension type and the inflection type. The first sees in scientific thought the shape of a decline common to secular thought in general that, at best, can point to its own failures (such as environmental catastrophe in ecology) and the second accommodates scientific thought as much as possible within the general bounds of its theology for the overall goal of bending or realigning those destructive aspects in both science and theology (such as anthropocentrism or chauvinism).

Chapter 2

Ecology and Thought

The most important concept within ecology is the ecosystem. This concept has changed over time, but the currently accepted definition is that the ecosystem is a physically locatable and quantifiable community formed by a system of energy exchange between the living, the dead, and the never-living where, when energy animates the system, there is an exchange of energy and material between the living and the dead.[1] It is not often understood that this concept, the most important and foundational within ecology, was created in response to the overdetermination of ecology by two schools of thought: organicism and mechanism. These schools of thought are philosophical and theological in their make-up and concern the reality of nature as such. The ecosystem concept is first articulated in 1935 by A. G. Tansley in his unification of these two rival schools of thought as they were understood and shaped by the material and stance of ecology. The organicists, primarily developed in and from the work of F. E. Clements, were opposed by the individualist reaction against organicism of Henry Gleason and his followers. In Clements's view the ecosystem, which he named "biome," was like a single organism where all the parts worked toward the health of the whole. Whereas Gleason rejected this organic view of nature and instead proposed that natural communities of plants are simply a random grouping of individual species that existed in that place because of the possibility of satisfying their needs, Tansley rejected the organicism of Clements, but could not follow the coincidentalism of Gleason, which constituted a decisive critique of Clements's views but did not provide any satisfactory understanding of the relation between plant communities. To overcome both their weaknesses Tansley forged a new theory from the dyad of holistic organic community and individualistic coincidental community.[2]

20 A NON-PHILOSOPHICAL THEORY OF NATURE

Tansley's conception of the ecosystem escaped overdetermination of ecology by certain philosophical schools, but he himself remained representationalist in his conception of an ecosystem. He considered the ecosystem a mental representation imposed on physical environments by the ecologist whereas the actual environment was a whole arising out of the prevailing relations. French ecologist Christian Lévêque explains it this way, "[The ecosystem] is thus an abstract reality formed mainly from elements that are themselves concrete."[3] In other words, Tansley's original conception of the ecosystem posits a nature that outstrips the ecosystem. Yet ecologists came to see that, by using the ecosystem concept as a way of organizing research and thereby advancing their methods, they could find concrete objects able to be identified in spatial and temporal terms that corresponded to the way the theoretical concept of an ecosystem worked. In other words, the ecosystem concept exists in both theory and as physical object, not just as mentally constructed. In a way, then, Tansley's original conception shows the continuing contamination of ecology by the nonecological. This is not uncommon as Daniel B. Botkin notes, ecology has often taken "an advance in scientific thinking at [its] time [...] and fixed it as though it were a permanent and final explanation."[4] The same holds true for ecology's interaction with philosophies and theologies of nature, as Botkin shows in his conceptual history of nature and ecology. In this section, following Botkin, we trace the interaction of ecology toward nonecological thought, like philosophy and theology, to show the shortcomings of these dogmatic images of nature imposed on scientific ecology and also to show that the current popular models do not refer to nature as such, but merely to the ecosystem. Ecology, much like environmental philosophy and theology, has also failed to think in an ecological nature, to mutate the philosophical and theological conception of nature with its own concept of the ecosystem. This is not an indictment of the scientific practices of ecology, however, as nature as such is not the object of research nor could it conceivably be the object of any single regional knowledge.

Looking through the history of ecological thought Botkin locates three "classical" images of nature and two that are "evolved," all of which were or are prevalent at different times and in different cultural milieus.[5] The classical or foundational images are Nature as Divine Order, Nature as Fellow Creature or as Organism, and Nature as Great Machine. These then evolve into the more sophisticated images of Nature as Computer or the Cybernetic model of Nature, and Nature as Biosphere. I will limit the presentation here to the broad characterizations of each image of thought just to highlight their particular determination of ecology. This

ECOLOGY AND THOUGHT   21

will stand in contrast to the attempt taken later to create a unified theory
of philosophical theology and ecology.[6]

Nature as Divine Order

Botkin begins his historical survey with Nature as Divine Order.[7] Under
this image, largely prior to the advent of ecological science, scientists
undertook the study of the environment under the presupposition that
nature tended toward an underlying equilibrium. This presupposition did
not rely on specific religious doctrines, but rather on what can be termed
"religious sentiment." Any discussion of dogmatic theology is notice-
ably absent from Botkin's overview.[8] Rather, in the scientific writing of
Christian scientists, Botkin locates a reliance on the thought of Cicero,
Seneca, Plato, and Xenophon.[9] Under this pregiven image scientists made
attempts to locate some sense of order (understood as equilibrium and pur-
posiveness) in nature, often in spite of observations to the contrary. This
belief in an ordered equilibrium led to an ecological theory of population
equilibrium between predator and prey, meaning that scientists assumed
there was a "natural" (meaning in actuality a truth that was transcendent
and eternal) relationship between predator and prey such that they would
always balance one another out. Botkin traced the failure of this view in
his own book through actual ecological case studies and one can see from
the practice of ecological restoration that the equilibrium theory simply
does not allow one to think the ecosystem as it really is. In other words, the
assumption of what is "natural" does not allow one to accept the actuality
or mode of existence of nature in ecosystems.

Botkin also highlights the anthropocentrism of this view. For the majority whose thought persisted under this image nature has been ordered specifically for human beings.[10] Botkin does not, however, discuss dominion
theory, which interprets this divine order as placing humanity at the pin-
nacle of creation as its Lord, or in more theologically respectable terms, its
Viceroy. Both views, however, present a very pernicious and ecologically
dangerous understanding of hierarchy. Everything praises the Creator of
the Divine Order through individual, dependent relations with the entities
directly higher up in the hierarchy. Thus, through the use made of nature
by human beings nature itself reaches its truest, best, and most beautiful
expression as given by its proper place in the hierarchy.

Under this image nature is ultimately only equilibrium, either as perfected or fallen, that is established correctly through hierarchy. Nature

22   A Non-Philosophical Theory of Nature

thus should be static and each being in nature should express its own static nature/essence if rightly attuned to the reality of the Divine Order as underlying nature/essence.

Nature as Fellow Creature or Superorganism

Botkin moves from the Divine Order to the organic image of nature.[11] This image takes a number of different shapes depending on the cultural milieu it develops within and is traced further back than the Judeo-Christian (one should also add Islamic, though Botkin does not) culture prevalent in the West today.[12] Common to all these varieties, however, is the view "that the Earth either is *like* a living creature or *is* a living creature."[13] Under this view the Earth is seen to be an idealized or perfected version of an organism with an analogous organic structure of growth and decay.

Ecologists and other scientists outside the field of ecology largely reject the organic view of nature today. It should also be noted that philosophers of the so-called naturalist-materialist orientation often mock this view, while many ecological philosophies and theologies have tried to reclaim it in the light of the "new physics." Despite this now ubiquitous position, there were scientific attempts to prove the organic view, most notably the attempts undertaken by the protoecologist and botanist F. E. Clements. Clements, like many pre-twentieth-century thinkers, looked for the implicit order underlying nature and saw there were associations of plants containing discrete individual plant species that would not persist outside the wider association of plants. This led Clements to the view that nature was like an organism made up of particular and necessary organs that needed to work as a whole. This was eventually disproven by Henry Gleason ultimately relegating Clements to a minor place within the history of ecology as a cautionary example of what not to think.

Botkin, like most ecologists, rejects the organic view of nature. He sums up the reasons succinctly saying that the organic view of nature was dependent, like the image of Divine Order, on a static understanding of plant associations, which was extended to more general associations of organisms, whereas observations show that individual species respond uniquely to environmental factors. Further, the species that dominates a particular environment changes continuously.[14] However, Botkin is obviously not comfortable simply mocking this view of nature. While he shows that this view is indeed wrong from the ecological standpoint, he is also convinced that the predominant mechanical view is also wrong and in the future both will be thought to be equally silly and misleading.[15]

Nature as Great Machine

Along with many other ecologists, Botkin clearly thinks that the mechanistic account of nature is wrong. Yet, it largely remains the dominant image under which ecology labors. He traces the outlines of this image last and emphasizes its connection with the image of Divine Order demonstrating the homology of the mechanistic and theocentric theory of the cosmos, as well its antagonism to the organic image.[16] Under this image of thought the Earth is dead, for it is a nonliving machine instead of a living organism.[17] The death of the Earth is predicated, however, on a theological perception of beauty: "The mechanical view is constant with the idea of a divine order in most of its particulars and consequences, and thus the mechanical perspective simultaneously reinforced the ideal of divine order and was reinforced by that theological perspective."[18] The theological perspective Botkin is here speaking of is, of course, the human search for a static and, due to that stasis, peaceful order to the cosmos that ultimately serves or can be manipulated to serve human ends. There is a deep connection here between an ecology guided by aesthetics and theological thinking:

[T]he belief in aesthetically pleasing and theologically satisfying physical symmetries was replaced by a belief in an aesthetically pleasing and theologically satisfying conceptual order. While the belief in gross physical attributes of symmetry, balance, and order was no longer tenable following the new observations of nature, Newton's laws created a conceptual order. Subsequently, theologians used this conceptual order to justify their belief in a perfect world where a perfect order (the laws of nature) ruled our asymmetric and structurally imperfect world.[19]

Thus the problem with the mechanical image is the same as the problem with the image of Divine Order as both attempt to locate simple, solid-state realities in nature despite the empirical findings of ecological fieldwork. Nature is not a great machine in the sense of an ideal, nineteenth-century machine that works according to an outdated physical model for the purposes of static predictions. The one truth that this image could give us has largely been occluded—nature is ultimately a duality of the artificial and the natural. From a certain perspective nature is, of course, natural but from another it is made up of artificialities as well. In other words, human beings can act as engineers or custodians of this machine for the benefit of all of nature (human and nonhuman). Yet, instead, the view has tended to see nature as a divinely constructed machine that must either be left completely undisturbed to remain perfect as such or completely subjected

24   A NON-PHILOSOPHICAL THEORY OF NATURE

to human mastery to be made perfect.[20] The mechanical image fails not because it displaces God, but because it perpetuates a theology that either negates one aspect of the earth, humanity, or kills the whole of the earth by refusing it life.

The Evolving Images: Cybernetic and Biosphere

Botkin then focuses his attention on two more images that are more advanced and sophisticated: Nature as Computer or the Cybernetic image and Nature as Biosphere or the earth as life-support system.[21] These images clearly have some origin in the preceding three, but make significant advances on them as they are revised by ecological findings.

The cybernetic image of nature, however, does not provide an adequate image of a nature that outstrips the biosphere or specific ecosystems, but using computer modeling guided by mature human observation and speculation can lead to helpful findings regarding particular ecosystems.[22] This is because the computer, unlike the ideal physical machine, can help ecologists to accurately model mathematically the stochastic reality of ecosystem development. The cybernetic model allows nature incarnated in ecosystems to be known as they are: nondetermined (i.e., nonmechanical) dynamic systems of complex exchange.[23]

This image can also be thought of alongside the other evolving image of nature: nature as biosphere or life-support system. It should be noted already that this image of nature is irreducibly coimplicated with the reality of the planet earth as such. Botkin provides a wonderful and grotesque description of this image by comparing it to the relationship between a moose and the bacteria that live in the moose's stomach and that the moose in turn needs in order to continue living in its particular environment.[24] The most well-known theorist of this image is, of course, James Lovelock who developed the famous Gaia hypothesis. Under this theory nature is thought of as a single system that regulates and makes possible the development and perpetuation of life. Thus, the biosphere metaphor says that there is some steady-state quality predicated for the planet as a whole, but not to individual organisms, and this steady-state quality must be understood within the wider stochastic events in large and small ecosystems.[25] It differs fundamentally from the organic image (though obviously owing it some debts), a fact often missed in environmental theology, in that the earth is not said to be alive, but is "a life-supporting and life-containing system with some organic qualities, more like a moose than a water-powered mill."[26]

ECOLOGY AND THOUGHT   25

Images of Nature and Images of Ecosystems

The first three images Botkin outlines are beliefs about nature as such. They propagated misunderstandings and became stratified as myths that require their continuing evolution into other images. On the failures of these images and the ecology that worked under their presuppositions, Botkin writes damningly: "Until the past decade ecology has remained a nineteenth-century science and has led us into failures in the management of natural resources and to unsettling contradictions in our beliefs about nature and therefore about ourselves."[27] The "evolving images" he traces in the proceeding chapters are more beneficial for those projects that aim to be intelligent and are considered in their ecological management of natural resources, our continued understanding of the earth, and our place in its systems, but they are not, properly speaking, images of nature. Instead, these are images of the abstract ecosystem. This is clear in Botkin's work as his use of the word "nature" changes from the three chapters on classical images to the two on evolving images. In the first three the attempt is to make sense of that which underlies particular natural areas and in the second we find an account of the successful descriptions of those natural areas, whether they are individual ecosystems or the world system of the biosphere.

In other words, in the first we see what Alain Badiou has called the passion for the Real.[28] This passion, as explained by Badiou, is an attempt to get beyond representation or semblance to what is truly real—to think the Real itself. This is precisely the motivation behind early ecological thought appealing to prior images of thought, to move beyond the semblance, beyond "mere empiricism" as we may be misled by our senses, to an unfolding of the science of ecology with the grain of the universe, in the light of what is truly real. In some sense, then, the early ecological thinkers recognized the perversity of nature, the ability of the Real to slip past any thought of it and this is witnessed in their appeal to philosophical and theological images of thought. That ecologists gradually stopped appealing to philosophy is not a sign that ecology's passion for the Real eventually faded, as passions often do, but was necessary because of the myriad errors that such images of thought were leading ecology into with regard to their primary material, the ecosystem. As shown in Botkin's switch in language regarding the "evolved" images of thought ecology in some sense had to bracket the question of nature, overdetermined as it was by philosophical and theological images of thought, in order to think the ecosystem as truly real without the semblance of philosophical Nature or as a representation of that Nature.

26   A Non-Philosophical Theory of Nature

Yet, this doesn't have to be the end of ecology and thought; this failure to bring together ecology and thought, that is, philosophy and theology, is limited because ecology was overdetermined by these images of thought. Ecology, perhaps for being somewhat more humble than the so-called thinkers of Big Science in biology and physics, has yet to go on the offensive against philosophy and theology; ecology has not tried to take the place of philosophy and theology, claiming that it can provide truth better than them.[29] Yet, the concepts that ecology has created and developed through its history of practice should be a challenge for thought, without thereby invading and colonizing thought, but instead being a posture that thought begins to think in or a kind of change in identity when one "goes under," both in the sense of going under medication for surgery and in the sense of going under a different name. We will return to this in the next chapter, but first we must turn to the relationship of philosophy and ecology, both philosophy of nature and environmental philosophy, to see what happens when philosophy refuses the challenge of ecology and the missed opportunity to make our thought ecological by "going under" ecology.

Chapter 3

Philosophy and Ecology

Many are looking to foster a relationship between ecology and philosophy as it becomes clear that the reality of our contemporary age, as well as the future that we are rushing headlong into, is determined in large part by the environmental crisis. This attempt is not unprecedented as the environmental movement and some form of environmental studies have been around at least since the writings of John Muir and Henry David Thoreau. The legacy of the relationship between ecology and philosophy has been and continues to be led by the discipline of environmental ethics and environmental aesthetics.[1] In this way philosophy prescribes ethical and aesthetic norms on the basis of facts given by scientific ecology, but philosophy itself tends to remain unchanged by the encounter. There may be some change, often favorable (a favorite is replacing the Western subordination of ethics to reason with principles from Eastern philosophy and religion), but what remains after this change is still a philosophical system, in this case based on ethics as first philosophy, developed apart from scientific ecology.

The relationship between environmental ethics and the science of ecology is quite clear: science cannot provide an ethics unto itself, only facts, which much be situated within some philosophical normative system that in itself is developed apart from ecology. This clean separation of fact and value is an instance of what Husserl called the "natural attitude," a phrase that already points to the complications philosophy runs into with this word—"nature." Yet, by bracketing the natural attitude, phenomenological thinkers have tried to think at a level more fundamental than the ought/ is distinction and thereby create a fuller environmental philosophy. Yet, the question remains: Is this form of thought ecological? What boundaries does it draw between itself and science? Phenomenological environmental philosophy still does not draw on scientific ecology out of fear of the natural attitude, but without asking if ecology unified with philosophy can provide a stronger thought outside of the natural attitude.

While these two strains of philosophical thought, mainstream environmental ethics and phenomenological environmental philosophy, are the dominant forms of philosophy and ecology there are some other resources for thinking about the relationship between the science of ecology and philosophy further afield. These thinkers do not call themselves environmental philosophers, but are instead concerned with nature as understood through the philosophy of nature. This resurgence of interest in the philosophy of nature has included a renewed philosophical interaction with physics or scientific cosmology and thus offers us another example of the relationship philosophy may take toward science in its attempt to think nature (understood as a first name of the Real). Here the problem of philosophy's relation to science remains. Does philosophy's understanding of nature change in an engagement with science or is nature still thought of primarily in a philosophical register? Furthermore, is there a theory of the relation between philosophy and a science other than physics?

Ethics and Ecology

Normally, owing in part to the "economics of thought," academic philosophical discourse operates via sharp distinctions between various philosophical domains. In order to be marketable the philosopher must advertise their areas of specialization and so one will either do "ethics" or "metaphysics" and in the interest of maintaining the distinctive character of their specialization they will impose this separation on reality itself as presented through their philosophical thought. Thus, you find in mainstream Anglophone philosophers, sometimes referred to by the anachronistic term "analytic philosophers," that ecology is still brought before philosophy and asked to reveal its ethical status. However, ecology presents certain challenges to these kinds of philosophical scissions, for the ontological status of an ecosystem and the ethical demand arising from its existence are not easily separable. Hence, Aldo Leopold's "land ethic" brings together the ontological and the ethical as Paul W. Taylor describes it, "The very structure and functioning of the Earth's ecosystems, it is said, make known to us the proper relationships that should hold between ourselves and the natural world. [...] The conclusion drawn from these considerations is that the science of ecology provides us with a model to follow in the domain of environmental ethics."[2] As a philosopher, though, Taylor rejects not just

PHILOSOPHY AND ECOLOGY 29

the specifics of Leopold's land ethic, but the very merging of the ethical and the ontological for breaking certain *philosophical* rules writing, "This line of reasoning is not sound from a logical point of view. It confuses fact and value, 'is' and 'ought.'"[3] He goes on to say that ecology, as a special branch of biology, can only provide facts that have to be taken into account when one philosophizes about the proper way to live within the physical limits for survival that those facts set. He goes so far as to claim, "Nothing in ecology, for example, can tell us that it is wrong to have a wholly *exploitative* attitude towards nature."[4] In other words, translated into terms familiar to readers of Heidegger, Taylor is making the claim that science does not think, in this case it does not think ethically. Philosophy must come alongside it and think for it ethically, even though, in the case of Leopold, you have a scientific thought and an ethical thought coming together in a unified manner, while in the case of Taylor's implicit metaphilosophy you have two modes of thought that are at war, or at the very least not at peace, with one another. The particular form this takes in Taylor is common to mainstream environmental philosophy insofar as it centers on certain philosophical problems, like the relationship of an "ought" to an "is," and takes these as problems that float around scientific ecology's relationship to ethics.

It is telling that in Taylor's development of a theory of environmental ethics he does not engage in a deep way with ecological concepts. Perhaps, one might suggest, Taylor constructs such a limit to philosophy and science owing to this philosophical commitment to a separation between ontology and ethics. Such a notion should be able to be tested by considering a philosophy that rejects this separation. One philosophy that appears not to suffer so acutely from this same separation of the ethical and the ontological and metaphysical is the deep ecology or ecosophy of Arne Naess, perhaps owing to the influence of Spinoza on his own work.[5] He recognizes that "one's ethics in environmental questions are based largely on how one sees reality" and thus holds that it is "important in the philosophy of environmentalism to *move from ethics to ontology and back*."[6] Yet, though this would seem to place it at odds with Taylor's strict separation of the ontological and the ethical, we find that both have the same limit-structure with regards to its practice and relationship to science. The notion of a deep ecology would appear to suggest that Naess's philosophy is developed alongside concepts from scientific ecology, yet Naess's real hope is to move from ecology to ecosophy. Ecology doesn't appear to set the agenda for the philosopher, but instead provides, as it so often does, a litany of facts about the destructive power of contemporary human society on the wider nonhuman world. Rather than challenge philosophy with ecology, as we aim, Naess provides philosophy and ecology with a

30   A NON-PHILOSOPHICAL THEORY OF NATURE

model that will be taken up by a plethora of environmental philosophers: Western philosophy, along with its complicity in the "so-called scientific worldview," is to be challenged with Eastern philosophy.[7] Naess doesn't do this in some naive sense, he isn't trading in a vulgar exoticism, but when it comes to concepts that he finds problematic in the Western philosophical tradition, such as the divide between objective and subjective qualities or the particular dominant form the concept of the self has, he draws on resources from the Eastern tradition combined with his own philosophical project rather than drawing on scientific ecology.

This is especially strange since he recognizes that ecology "has application to and overlaps with the problems of philosophy."[8] So what is it that keeps Naess from engaging deeply with ecological concepts? The answer is that ecology as a science is suspect precisely because it is a *science*; it operates with the suffix -logy rather than -sophy. Within Naess's ecosophy science must controlled, including ecology, science must be placed within a normative, that is, philosophical, milieu that limits its power, or as Naess would rather say, that recognizes the limits of its power. The impetus behind this ecosophical reigning in of science is similar to an axiom that I am working from; namely, that Real, or nature as a first name, resists any total capture. Naess recognizes this resistance in his own way but unlike our project uses it to critique the post-Galilean scientific worldview, claiming that we must resist any kind of universalization of one science, be it biologism or ecologism, which also serves to weaken the science's internal identity by generalizing its concepts too much.[9] In part this is because, for Naess, scientific ecology like all natural sciences engenders an understanding of nature that is too disjointed. He even goes on to claim that such a relationship between philosophy and ecology would favor a shallow ecological movement.[10]

What then are the limits to scientific ecology, which Naess thinks are necessary to engender a "profound" understanding of nature that undergirds a deep ecology? Against ecologism, the overgeneralization of concepts from ecology understood simplistically, we can locate an ecological minimalism at work in Naess's ecosophy. We may even call this minimalism shallow ecologism, as it refuses a deep engagement with scientific ecology. This is operative in Naess's definition of ecology: "The expression 'ecology' is infused with many meanings. Here, it will mean the interdisciplinary scientific study of the living conditions of organism in interaction with each other and with the surroundings, organic as well as inorganic. For these surroundings the terms 'milieu' and 'environment' will be used nearly interchangeably."[11] This isn't a bad definition of ecology; in fact, it is quite close to the generally accepted definition given in Michael Allaby's *A Dictionary of Ecology*, which states, "The scientific

PHILOSOPHY AND ECOLOGY   31

study of the inter-relationships among organisms and between organisms, and between them and all aspects, living and non-living, of their environment." However, it doesn't delve into ecology's concepts with any depth either. Even the concepts it touches upon, "organisms" (populations, or the diversity of species that populate the ecosystem), "living conditions" (what we call the never-living space and temporality of the environment), "interaction" (energy relations of exchange that arise out of the populations interaction with one another), and "environment/milieu" (ecosystem), are not explored in any depth in relation to philosophical issues.

This lack of depth is likely because Naess thinks that the science of ecology only provides us with a recognition of our severely limited ecological knowledge, and ecology tells us that we don't yet understand the ecological consequences of change in a particular ecosystem: "The study of ecosystems makes us conscious of our ignorance."[12] Indeed, the only truly positive notion that Naess appears to take from ecology is the idea that "all things hang together," which he takes to be an ontological statement that is ethically significant.[13] Yet, Naess points out that this does not in itself explain how all things hang together, but instead of turning to the very things that ecology is precisely not ignorant about he turns to another philosophy (Gestalt thinking).[14]

But ecology could provide resources for understanding how things hang together because it is not ignorant of the aspects mentioned earlier (biodiversity, energy exchange, the spatial borders of an ecosystem, etc.); ecology has developed a number of tools for understanding the various ecosystems and the wider biosphere. So what exactly is Naess referring to by claiming that ecology reveals our ignorance? Naess says that this has to do with a kind of political usefulness. No longer can politicians appeal to science or instrumental reason to deal with pressing issues; cost-benefit analyses will no longer be a substitute for wisdom.[15] If this were the case it would be laudable, but there is a slightly more nefarious effect of Naess's presentation of ecology, one that we will see mirrored in other philosophical thinkers as well. In short, Naess is claiming like Taylor, as discussed earlier, that science, and ecology specifically, does not think. That only philosophy, in the guise of ecosophy, can provide the framework to make any practical sense of the statements of ecology. In a statement remarkably similar to the problematic position of Taylor outlined, Naess writes, "Without an ecosophy, ecology can provide no principles for acting, no motive for political and individual efforts."[16] Again we are left with a split in reality between what we know *is* and what we think we *ought* to do.

In a strange way, then, Naess is actually accepting the arrangement of philosophy and science that is unecological within thought. Not only does Naess not draw on scientific ecology to challenge and push philosophy

32   A Non-Philosophical Theory of Nature

on problems inherent to it and related to environmental issues, he also doesn't attempt to mutate directly what he takes to be science's underlying *philosophical* split between primary and secondary qualities and objective and subjective reality. Rather, he continues the typical relationship of science and philosophy: philosophy over a science that does not think. At one point referring to the scientific study of the environment he asks sarcastically, "Are we getting any closer with the long scientific strides built upon the work of Galileo or Newton?"[17] If the goal of a deep ecology movement was to turn the tide of environmental destruction by fostering an ecosophy, can we not turn this question back on Naess? Are we getting any closer to an ecosophical relationship with the biosphere with the strides built upon the work of Naess? While, as with all the philosophers outlined in this chapter, there is much of value in Naess's work, the relationship of philosophy and ecology remains in itself unecological, even unecosophical, and that means that the very form of his thought remains largely trapped in the self-sufficiency of a philosophical form of thinking that sets itself up to remain unchallenged by the form of ecological thought.

Phenomenology and Ecology

Naess recognizes that a true ecosophy cannot separate the ethical and the ontological as strictly as Taylor, though he organizes ecosophy along a similar structure such that the science of ecology is, like an "ought," unable to provide any principles for action. We find the same general structure of science requiring a philosophy purified of its scientific deviations at work in the phenomenological tradition, while providing a philosophy that is, ironically given Naess's aims, more "lived" or "everyday" than ecosophy. Rather than proceeding from norms, that is, from what "ought," phenomenology proceeds from the thing itself, from the essence of what appears to us because these norms are already suspect as developing within the natural attitude. The natural attitude names that particular way of thinking that we take as given, but that, if we begin to think critically about the structures of our thought, appears instead as just a frame of reference or meaning within a wider pregiven horizon that Husserl calls the life-world. In this section I'll consider the underlying structure of phenomenology's relationship to science as it is developed in Husserl, which is defined by the "crisis of the sciences," and how that is developed by the environmental phenomenology of John Llewelyn. What will be common to all, though, is the submission of science to phenomenology, to a form of philosophy developed necessarily apart from science.

PHILOSOPHY AND ECOLOGY   33

In the contemporary world of philosophy many hold the opinion that phenomenology's relationship to science was and remains antagonistic, that it seeks to undercut the power of science and/or that it ultimately fosters a nefarious and pathetic agnosticism in its rejection of classical metaphysics.[18] In reality, though, this harsh view of phenomenology ignores the nuance of phenomenology's relationship to science, at least as is found in the writings of its founder Edmund Husserl. Husserl rejected a certain philosophy of science that he saw at work in European human society. This he called the positivistic reduction of science, where science merely provided facts. This philosophy of science should sound familiar as we found it at work in our survey of mainstream environmental ethics mentioned earlier. This limiting of the meaning of science to the "factual sciences," or rather simply to those aspects of the sciences that are concerned with uncovering facts, is what is of concern for Husserl. His engagement with science, then, "concerns not the scientific character of the sciences but rather what they, or what science in general, had meant and could mean for human existence."[19] The crisis of the sciences, then, is actually a crisis of human existence because "[m]erely fact-minded sciences make merely fact-minded people."[20] While the nihilist philosophy of Ray Brassier, which boldly declares that we are "already dead," may still have a problem with Husserl's attempt to subsume science into the realm of philosophical meaning, Husserl's philosophy does not take aim at science as such but at the unacknowledged philosophical commitments at work in people's understanding of science, what we may also call ideology.[21] No, the problem is not that Husserl fears science or can't bring himself to reconcile his beliefs with some empirical facts that, we are told, stand opposed to his beliefs, but rather that his philosophy remains untouched by concepts derived from science. He still sets up a hard boundary between what science says in its everyday practice, what its axiomatic concepts are, and the meaning that philosophy can provide, which can hedge in what we now call "Big Science."

Husserl's claim is that the life-world is "dressed up" in the notions of mathematics that are absolutized, and though this leads to discoveries Husserl considers important, these notions ultimately confuse "true being [for] what is actually a method."[22] What science requires, because science is in crisis, according to Husserl, is a philosophy that remembers the life-world, which is its "meaning-fundament."[23] That is to say that the natural sciences remain naive without any kind of fundamental inquiry into the very life of things, without questioning that there is in fact a life of things, or in Heidegger's terms, without a fundamental inquiry into what the Being of beings is, what thinking is, and so on.[24] The natural sciences, Husserl claims, need to be returned to this life-world or risk losing

34   A Non-Philosophical Theory of Nature

it; indeed, losing life itself. Husserl's solution to this problem is locating an anonymous, transcendental subjectivity that is "functioning in all experiencing, all thinking, all life, thus everywhere inseparably involved," but that itself has "never been grasped and understood."[25] How he gets to that transcendental subjectivity, however, is what is ingenious in Husserl. Notice that he accuses science divorced from the life-world as being naive.[26] Yet, the solution to that naivety is to plunge into it intentionally, whereas before one simply acted in it. This may become clearer if one considers this in light of Plato's familiar cave myth. There we have the prisoners, chained to a wall since birth and made to watch shadows of people, animals, and the like dance on the wall of the cave. This is their only frame of reference so that they take, completely naturally, these shadows as truth. When one of the prisoners escapes, whether through accident or intention, and emerges into the "real world," he begins to see things as they really are or, at least, as *more* real than they are in the cave. Husserl, though, sees no reason to leave the cave. In fact, we have every reason to question the notion that outside the cave is the "real world." What is outside the cave is just the world beyond the cave, the cave itself is part of the real world, as are the materials in the cave that hold the prisoners to the wall and the materials for projecting the shadows upon the wall. No, what the usual telling of Plato's myth serves to do is provide a cover for a more insidious cave. If we consider this in the Hollywood terms of the twentieth-century's film version of this myth, Lana and Andy Wachowski's *The Matrix*, it would be as if there was a matrix inside the matrix. So when Neo takes the red pill he simply enters into another version of the matrix, one that chains him ever more for his having thought that he escaped illusion. Indeed, this set of problems is presented cinematically in the 2010 film *Inception* directed by Christopher Nolan.

Husserl's radical step is to perform an epochē, what is also called the reduction or bracketing, on what appears. Eugen Fink, one of Husserl's closest students and an assistant on many of Husserl's most important projects, including the *Crisis*, describes the epochē as a radical new beginning for thought. "The *de-absolutizing of the world* (which in the natural attitude is *absolutized*) signifies a more radical 'Copernican revolution' than the conversion from a geocentric to a heliocentric system—one more radical than all the philosophical revolution in world outlook which takes place on the basis of the natural attitude."[27] There is something radical within Husserl's thought, a radicality that is perhaps not available to us now after the institutionalization of phenomenology. Yet, what Husserl's epochē did was take away the whole world from us, though of course he did so in order to bring it back, but bring it back under new conditions, as a phenomenon constituted by an anonymous transcendental subjectivity.

PHILOSOPHY AND ECOLOGY   35

There is a two-step approach to the radical or transcendental epochē. First
there is the epochē of objective science, bracketing all the inherent philosophical notions that plague scientific thought: "What is meant is rather
an epochē of all participation in the cognitions of the objective sciences,
an epochē of any critical position-taking which is interested in their truth
or falsity, even any position on their guiding idea of an objective knowledge of the world."[28] This, though, is not enough for Husserl. It retains
too much still of the natural attitude; those opinions that persist outside
science and that may be more pernicious than the quest for objectivity.
"What is required, then, is a *total* transformation of attitude, a *completely
unique, universal epochē.*"[29]

Fink describes this aspect of the reduction, which takes thought down
to its meontic absolute—bare life itself:

> The self-reflection of the phenomenological reduction is not a radicality that
> is within human reach; it does not lie at all within the horizon of human
> possibilities. Rather, in the actualizing of the reduction a self-reflection
> occurs that has a wholly new kind of structure: it is not that man reflectively
> thinks about himself, but rather that transcendental subjectivity, concealed
> in self-objectivation as man, reflectively thinks about itself, beginning *seem-
> ingly* as man, annulling itself as man, and taking itself down as man all the
> way to the ground, namely, down to the innermost ground of its life.[30]

Where we stand, then, in this transcendental epochē is above the world,
above the validity of the pregivenness of the world.[31] This transcendental
stance is above the flux of the world, above the subjective-individual consciousness and intersubjective consciousness. This "unnatural attitude,"
transcendental to the world, bestows on the philosopher a position above
worldly interest:

> Any interest in the being, actuality, or nonbeing of the world, i.e., any
> interest theoretically oriented toward knowledge of the world, and even any
> interest which is practical in the usual sense, with its dependence on the
> presuppositions of its situation truths, is forbidden; this applies not only to
> the pursuit, for ourselves, of our own interests (we who are philosophizing)
> but also to any participation in the interests of our fellow men—for in this
> case we would still be interested indirectly in existing actuality.[32]

In a certain way the epochē is but a deepening of the scientific approach to
thinking. The description of the philosopher who has undergone this transcendental epochē is not far off from the description of the scientist unconcerned with the consequences of his actions for the rest of humanity, he
simply wants to know. Think of the scientists involved in the Manhattan

36   A Non-Philosophical Theory of Nature

project, who did not know what the effect of the atomic bomb would be, but who went out to the desert, put on the their goggles, and detonated it to find out. They did this knowing that one possible scenario could have been the complete destruction of the atmosphere, meaning the complete anni- hilation of all life on earth. But, though Husserl's description is far from the nostalgic gatekeeper of meaning that he is presented as by Brassier and others, it does share certain less dramatic qualities with science. Consider again ecology, where after the failures of chaining ecological science to pre-ecological philosophical images of natures, it tended to consider the ecosystem simply as it appears. Instead of approaching the ecosystem with an aura of metaphysical presuppositions, presuppositions that unlike the physicist the ecologist isn't allowed because of the inherent complexity of an actual object, the ecologist approaches the ecosystem naively with only a handful of axiomatic concepts that can be revised on the basis of the "presentation" of the object. Husserl's philosophy attempted to do the same thing within the field of philosophy. Throwing off the failures of meta- physics, not to encourage some agnosticism, but to truly engage the reality of things. Our goal, as it has been throughout this chapter, is not to assess the success of that attempt, but only to trace philosophy's self-constructed limits with regard to science. In that regard it is telling that Husserl's dis- cussion of transcendental subjectivity, a life that runs through things, has nothing to do with the way that science thinks life. In fact, science is now treated on this subject antagonistically in the *Crisis*. We can see this clearly when Husserl writes:

> The radical consideration of the world is the systematic and purely internal consideration of subjectivity which "expresses" itself in the exterior. It is like the unity of a living organism, which one can certainly consider and dissect from the outside but which one can understand only if one goes back to its hidden roots and systematically pursues the life which, in all its accomplishments, is in them and strives upward from them, shaping from within.[33]

At the far end of the phenomenological tradition this line of thought ends with the philosophy of Michel Henry, who has no room at all for sci- ence in his philosophy and instead delivers his own condemnation of the Galilean sciences writing, "In its inaugural decision, having placed sensible life, phenomenological life in general outside its field of study, Galilean science would assuredly not be able to discover it again through research, even though it calls itself biology."[34] After quoting the words of Nobel laureate François Jacob, "Biologists no longer study life today," Henry goes

PHILOSOPHY AND ECOLOGY   37

on to declaim, "We must take [biology] at its word: *in biology there is not life; there are only algorithms.*"[35] In distinction to Henry, Husserl spends most of his *Crisis* examining the philosophy of objectivism that philosophers fall prey to, rather than engaging in polemic against the sciences as such. The notion of being "scientific" for philosophers is the equivalent of taking the red pill, falling into a second matrix, but falling deeper into illusion for thinking you have escaped it: "Thus nowhere is the temptation so great to slide into logical aporetics and disputation, priding oneself on one's scientific discipline, while the actual substratum of the work, the phenomena themselves, is forever lost from view."[36]

While clearly for Husserl the separation of life from biological life, or science from the life-world, is not an error or failure of science as such, but rather one of a certain philosophical way of thinking that pervades science, we still see in the block quote earlier a skepticism toward science providing any specific tools for understanding this transcendental subjectivity. How Husserl submits science to philosophy is far more interesting than those philosophies discussed earlier, even if the general structure remains the same. Husserl takes a certain scientific attitude toward the world, naivety, and radicalizes it. We must still come before the world with our metaphysical presuppositions bracketed, and even more radically bracket the whole notion of world; so must science in such a way that it must be placed within a new philosophical milieu before it can operate outside of a permanent crisis. That is, only as a philosopher under a transcendental epochē can one say with Husserl, "I stand *above* the world, which has now become for me, in a quite peculiar sense a *phenomenon*."[37] In this way, science is both prized and distrusted. How, though, does this work itself out in relation to ecology?

Despite phenomenology's abiding interest in nature, it would be unfair to expect an explicitly ecologically informed phenomenology from thinkers like Husserl or Heidegger. Now in the age of ecological crisis, however, it is unsurprising to find phenomenologists engaging environmental problems. Yet, of the many books published, few environmental philosophers take up the radical stance of Husserl's return to the things themselves but instead produce studies of past phenomenologists' considerations of nature and the environment, sometimes putting them into their historical context and sometimes suggesting aspects that might be of use to addressing our current situation.[38] One notable exception is the Welsh phenomenologist John Llewelyn, whose two books of ecophenomenology, *The Middle Voice of Ecological Conscience* (1991) and *Seeing through God: A Geophenomenology* (2004), remain the deepest phenomenological engagement with environmental thought.[39] In both books Llewelyn takes this radical notion of a

38 A NON-PHILOSOPHICAL THEORY OF NATURE

transcendental subjectivity, as ethically reworked by Emmanuel Levinas, and considers how it challenges certain utilitarian environmental ethics, like those advocated by Peter Singer, and what this stance above worldly interests would mean for environmental philosophy. In so doing Llewelyn argues that we need to begin looking for a "deeper and wider ecology" alongside a "morally deep contractualism"[40] In other words, if utilitarian ethics are built on a strong sense of self and self-interest, rationally understood to mean that one's self-interest is always in some sense intersubjective, Llewelyn suggests that there is instead a responsibility toward all that is regardless of what they are.[41] While it may not, at first glance, appear this way, what Llewelyn has done is to put the seemingly morally monstrous position of the scientist, willingly to risk all worldly reality to know, within an ethical framework. The transcendental subject of phenomenology is not limited to a human being, it is the transcendental subjectivity that runs through all that is and it is from this position that one's worldly interests slip away, that one's ego is deabsolutized so that an absolute ego may be recognized.

Llewelyn interestingly connects this to James Lovelock's often-misunderstood Gaia hypothesis. In his reading of Lovelock, better than most, Llewelyn describes how the Gaia hypothesis performs a similar deabsolutization of the human world; the world understood through the natural attitude, writing, "the hypothesis [that Gaia will look after herself] bodes well for life in general. But the influence of the biosphere on the atmosphere is not necessarily good news for life in its human manifestation. That thought has at least the salutary effect of leading human beings to adopt a less anthropocentric perspective."[42] This is an interesting mutation of phenomenology with quasi-ecological material (the status of the Gaia hypothesis in the ecological literature is fraught, to say the least, though some acceptance of its general ideas of coevolution has been appropriated) and it could be helpful in a further complication of philosophy with ecological material.[43] However, throughout both *Middle Voice* and *Seeing through God*, this is the only scientific concept engaged with in any depth and that engagement happens without any wider theoretical consideration of the relationship between an environmental phenomenology and scientific ecology. So even the most successful engagement with ecology, at least on our terms, still operates without any deep complication of philosophy with ecology. While Husserl clearly meant that there was a crisis in the whole of European life, this, along with Heidegger's own criticism of the drive toward objectivity in European science, nonetheless, created a posture of antiscience within phenomenology that persists in environmental phenomenology. Phenomenology, then, remains unecological in thought.

Philosophies of Nature and
Blindness to Ecology

Among many contemporary philosophers there has been a renewed interest in the philosophy of nature. Unlike the previous forms of philosophy surveyed earlier these philosophers of nature are largely unconcerned with creating any kind of environmental philosophy. We are looking to them, though, because the question that drives this work is not limited to the sphere of environmental thought, dominated as it is by questions of ethics, but by the question of an ecological nature, to think nature in an ecological way that, we hold, has not been done before. Philosophers of nature, then, offer an alternative model of thinking nature whose boundary between science and itself will be different than environmental philosophy. Thus, in this section, we will trace a different form of the limit of science and philosophy than the previous section, a form that may prove a better model for science and philosophy, but whose failure to be able to think the ecosystem is demonstrative of our claim that nature is not hidden, but perverse. That nature cannot be grasped by a single thought, but instead determines every thought. Again, as with the surveys earlier, this is not intended to be a comprehensive assessment of the philosophy, but only intended to trace its limit experience with regard to science and, in this particular case, to locate in philosophy some resistance to the subsumption model, meaning that science can only think if it is subsumed into philosophy, which is common to environmental ethics, deep ecology, and phenomenological environmental philosophy. In the case of these new philosophies of nature this resistance is accomplished without thereby becoming a full-blown scientism.

This renewal of interest in the philosophy of nature is in part a reaction to the twentieth-century critique of metaphysics common in both the analytic and Continental philosophical worlds. A variety of speculative thinkers, such as Ray Brassier and Quentin Meillassoux, already mentioned earlier, but also Iain Hamilton Grant and some others, have captured the imagination of many younger philosophers and theorists in their call to return to the "great outdoors."[44] In each case these thinkers engage strongly with the natural sciences: mathematics and cosmology for Meillassoux, neuroscience for Brassier, and physics for Grant. Unlike Husserl, these philosophers are unconcerned with any crisis in science, but instead take science as necessary for philosophical work.

Of these thinkers it is Grant that is concerned most explicitly with the philosophical problem of nature. His work proceeds much like Deleuze's early history of philosophy texts, entering into the discourse of a past

40   A Non-Philosophical Theory of Nature

philosopher and pushing it to its limits. For Grant that philosopher is
Schelling, though he engages with the entire constellation of thinkers
known as the German Idealists. The whole of the argument concerning nature can actually be seen in the difference between the Fichtean
and Schellingian systems of philosophy for Grant.[45] For Fichte's understanding of nature, containing with it a rejection of nonliving nature, the
splitting of nature and freedom into two different "worlds," is the common view of nature among philosophers, which Grant claims have been
practicing an antiphysics for some two hundred years.[46] Against Fichte's
notion that nature is to be understood primarily through animal being,
that is, as an organism, such that "life exhausts nature," Schelling makes
the argument that nature is also partly living.[47] Grant locates Fichte's error
in the relationship between philosophy and science, while noting that it
was Schelling's deep engagement with the speculative physics of his age
that allowed him a richer and more consistent philosophy of nature. For
"Fichte's only error [...] is to deny activity to nature on transcendental
grounds while rejecting the central precept of dynamic physics [...] i.e.,
that there is no substance behind the powers."[48] Clearly, for Grant following Schelling, the transcendental cannot think alone, for the transcendental is itself constituted by nature, and thus philosophy will always be
bonded to *the* science of nature, which he claims is physics.

Now, we can begin to see the problem within this model of science
and philosophy if we consider perhaps the most influential contemporary
Schellingian thinker of nature, Slavoj Žižek, who uses Schelling's philosophy of nature along with contemporary discussions of physics to cast a
vision of nature that is ultimately "not-whole."[49] A nonharmonious vision
of nature that runs afoul of the usual portrayal of nature that, Žižek claims,
is found in environmental thought. Setting aside the fact that Žižek's portrayal of environmental thought appears to be pitched more for contrar-
ian affect than any interesting philosophical purpose, his understanding
of nature via Schelling and physics that is then turned against ecology
as such, which we must also understand as scientific ecology despite his
lack of specificity, is telling of the inherent blindness of these philosophies that take themselves as eternally bonded to physics. For Žižek claims
to be speaking about nature, and as support for his view he appeals to
the philosophy of Schelling, already outlined earlier, and to findings in
contemporary physics.[50] Yet, he deploys this against ecology as if ecology
were not itself another science of nature. The incompleteness of nature is
not taken as a challenge to engage with multiple sciences of nature, but
instead taken as a sign that science has caught up to the most brilliant of
philosophies: "Because of this self-reflective character of [quantum physics'] propositions, [it] joins ranks with Marxism and psychoanalysis as one

PHILOSOPHY AND ECOLOGY   41

of the three types of knowledge which conceives itself not as a neutral adequate description of its object but as a direct intervention in it."[51] This is, of course, ridiculous since ecology, in order to describe its object accurately, must account for their own direct engagement with nature and it is impossible to avoid some intervention in an ecosystem. But, more importantly, while Žižek and other philosophers of nature following Schelling do not subsume science into their philosophy, their raising of physics alone to philosophy's level still constitutes blindness to the perversity of nature.

Of course, Grant does not share this particularly distasteful aspect of Žižek's philosophy of nature. In fact, he provides us with a powerful thought-experiment that demonstrates the perversity of nature—a thought-experiment called the "extensity test."[52] For Grant, following Plato and Schelling, sees philosophy as the universal science and as such nothing is outside its remit, it must be bold enough to think the All. Like Schelling he aims to think what philosophy has left unthought, nature, the unconditioned of thought itself:

[Philosophy] is "the infinite science," and cannot therefore be "conditioned" by eliminating anything a priori from its remit [...]. The infinite science must test itself against the All, which lacks neither nature nor Idea. It is the extensity therefore, the range and capacity of philosophical systems that is being tested [...]. [Schelling] challenges systems to reveal what they eliminate. Insofar as philosophy still leaves nature to the sciences, it continues to fail Schelling's test, and becomes a conditioned, that is, a compromised antiphysics.[53]

Insofar as nature lies outside the remit of just one science, can a philosophy of nature ever hope to think nature if it is eternally and necessarily bonded to but one science, the science of physics? Of course, Grant is not suggesting that physics as such is the final arbiter of the truth of nature, indeed his argument is that by separating physics from philosophy it has failed to live up to its own identity as the infinite science. That in separating the subject of physics, *physis*, from philosophy it has left itself blindly conditioned by nature. A philosophy that is a compromised antiphysics will always find itself coming up against the chiding of facile naturalism, and what lies behind the laughter of these naturalist philosophers, who are of course compromised as well, is what the compromised antiphysics has left unthought.

The point here is not to say that Grant has left ecology unthought, which would be rather silly since his work thus far has been a historical study to set up his own forthcoming project, but to challenge the philosophical overdetermination of nature, both as an idea and as that which is (though,

42   A NON-PHILOSOPHICAL THEORY OF NATURE

in-the-last-instance, these are both nature as One). Grant's point, following
Schelling, is not a return to physics as such, but to the subject of physics,
physis/nature, a return to what underlies a priori the conditioned within
thought. As Grant says concerning Schelling, though this is equally true
of most philosophers involved in the return to nature, "Schelling always
'starts with' naturephilosophy because 'nature IS *a priori*.'"[54] Yet, there is a
temptation at work in the recent turn again to nature to think that a single
science offers us access to some essence of nature that may then be enriched
through a philosophical operation or, as with Brassier, provide the thought
through which all philosophy must be judged. In Grant's work this plays
out as a development of a philosophy of forces, obviously derived from
physics but rethought within philosophical problematics. Yet, a philosophy
of forces fails the extensity test when it comes to problems related to ecology; it simply is too limited to think the complexity of ecological issues
of ecosystem organization, biodiversity, niche construction, and energy
flow.[55] This is clear from Žižek's own engagement between a philosophy
infused with physics, as ecology is denied even access to nature, but this
holds true also with regard to a philosophy that derives only forces from
nature and thinks from there. While that is likely to be valuable, certainly
the challenge put forward to philosophy, that nature exists for it, needs to
be meet by a thousand unified theories of philosophy and science X, rather
than any kind of subsumption of some other science to a unified theory of
philosophy and physics. After all, any attempt to do so will ultimately lead
to failure, for as E. O. Wilson remarks, "Physicists can chart the behavior
of a single particle; they can predict with confidence the interaction of two
particles; they begin to lose it at three and above. Keep in mind that ecology is a far more complex subject than physics."[56]

The underlying issue here concerns the status of particular sciences,
like physics or ecology, in relation to philosophy. While the limit between
science and philosophy is clear in the approaches outlined earlier, what we
have called the subsumption model, within new philosophies of religion
there is more room for a sharing of material. A model that I am fundamentally in agreement with, but that remains, in my view, unstable because the
idea of science itself remains unthought. In the next chapter, building on
the work of Laruelle, I present a model and theory of science that moves
beyond these problems. But, while thinking through the relationship of
physics to philosophy, the question of physics overdetermination of the
idea of science is raised. Corollary to that question is the need for any bond
between philosophy and ecology, the question of whether or not ecology
is a science.

Lévêque points out that "in the eyes of some people and decision-makers,
and even in the eyes of scientists, ecology is not generally viewed as a

PHILOSOPHY AND ECOLOGY   43

science in the same way as nuclear physics or molecular biology."[57] Some of these doubts have been founded, for as with any science ecology has had to respond to its own failures and overdeterminations by unscientific thought, some of which we outlined in our discussion of ecology and thought. Yet, the overwhelming reason that ecology is often derided as unscientific is its inability to test its main concepts and to provide hard laws for ecosystem functions, at least laws as defined by physics and chemistry. While there is an absence of laws, there are, as Wilson says, "as in the study of evolution [...] principles that can be written in the form of rules or statistical trends."[58] There can be no hard laws in ecology owing to the complexity of its object. We can only begin to understand such a complex and perverse object through heuristic principles, or, in other words, through practical axioms. Thus, the main concepts of ecology, which are not just concepts but actual realities or propositions, cannot be falsified in the sense of Popper's philosophy of science. Yet, like Žižek's quantum physics, this is in part because the reality of ecology, that is, a certain earthly manifestation of nature, demands that the observer is somehow implicated in the empirical findings.[59]

While Lévêque's hope for a mechanistic account of the ecosystem sometimes leaks out in his account of ecology's history, and thus also the edges of a certain faith in mechanistic philosophy peek out as well, the likelihood of that account sharing much in common with the model of physics is low. Not because scientists are conspiring against beauty and freedom—that idea is largely dependent on one's preunderstanding of beauty and freedom and has little to do with mechanics as such—but simply because ecosystems are neither chaotic nor ordered enough to fit such an abstract understanding of nature. Consider Wilson's account of ecosystems. He invites us to think of two different extreme models for how an ecosystem may be organized with regard to its biodiversity. The first is total disorder; here species come and go without any strong relationship with the presence or absence of other species. The second is perfect order; here the ecosystem itself is just a superorganism and thus if we could just know one species within the ecosystem we would be able to understand the whole of the ecosystem without further study of individual populations.[60] As remarked in the earlier discussion of ecology, to be developed more later, ecologists reject either extreme as a principle for ecological science. Instead, "[t]hey envision an intermediate form of community organization, something like this: whether a particular species occurs in a given suitable habitat is largely due to chance, but for most organisms the chance is strongly affected—the dice are loaded—by the identity of the species already present."[61] In short, the manifestation of nature that ecologists study and explain is a complex system containing actualities rather than abstractions and as such it

44   A NON-PHILOSOPHICAL THEORY OF NATURE

requires axioms and heuristic concepts, but philosophers of nature should not run away from complexity least they leave it unthought and, perhaps more importantly, leave philosophy unchanged by complexity.

Expanding to consider nature in its complexity, using concepts that have been successful in the scientific study of ecosystems within philosophical problems, as well as on the structure of philosophy itself, is surely a step forward in terms of meeting the extensity test. For if philosophy has been bonded eternally and necessarily to physics, would not the philosophical conception of physics, as a philosophy of science, determine its understanding of nature and in turn lead scientific thought aground if this conception is misguided? If, after the dominance of Popper's philosophy of science, we are beginning to understand that some forms of science are structurally unable to fit the paradigm of physics, must not the paradigm of science itself change? We will return to this question in the next chapter. For now, though, we have demonstrated the perversity of nature with regard to philosophy's attempted capture, either through the subsumption model or through the bonded model. Wilson says of ecologists that they "like the organisms they study, cannot make nature conform to their perfect liking."[62] This is a lesson for philosophy, one that it would be wise to take under consideration. In contrast to the structural weaknesses in the relationship between science and philosophy and philosophy's attempts to think environmentally, ecology may provide a model for how to think from nature, rather than from a hallucinatory conformism, but more importantly how to think nature in an ecological way.

Chapter 4

Theology and Ecology

Writing about theology in relation to ecology may seem strange within the scope of a book that itself stands within no particular religious tradition, or, more exactly, is unconcerned with its standing within any particular religious tradition as it aims for a "secularity of thought" within its own "non-theological" discourse.[1] There may even seem to be an air of insincerity or a cynical appeal to the power of a large group present in religious communities, a group that one may disagree with because they "don't know any better" but that can be exploited for the same reason, an attitude present in the appeal of E. O. Wilson to religious believers in his *The Creation* (2006).[2] There is however an ultimately realist reason for considering theology within the wider paradigm of "environmental thought": the form its thought takes is superior in a specific way to that of philosophy. Theology is concerned with many of the same questions as philosophy, and that includes the question of ethics and metaphysics raised by the ecological crisis. However, unlike the environmental philosophies discussed earlier, the works of ecotheology tend not to separate nearly as easily the ethical and the metaphysical, even within their own self-understanding of the discipline; these two strands are always unified within some broader paradigm, some single, theological vision. As will become apparent in the next chapter, this prefigures the methodology of non-philosophy; it is akin to the "vision-in-One" that non-philosophy thinks *from*. Yet, theology, I will argue, does not itself break with its own self-sufficiency, it does not allow itself to be thought ecologically. This is clearest in its interaction with the science of ecology.

In chapter 1, I say that the limit of theology's relationship to science is determined by the absolute subsumption of science into theology, though this subsumption operates differently than philosophical subsumption

46   A Non-Philosophical Theory of Nature

insofar as theology claims to subsume itself into the same vision-in-God that it would plunge science into. Or, in other words, theology operates by claiming a certain kind of poverty in its operation; a dependence on God, which is sometimes radical and sometimes relative depending on the theologians' particular understanding of God and World, but importantly God is never captured by theology (we will return to our understanding of theology in the next chapter). Therefore within theology we can differentiate two different types according to the operation of this subsumption: declension or inflection. The declension type is to be understood as distinct from what is more commonly referred to as the declension narrative, or rather, the declension narrative exists within a wider posture of declension toward thought. The narrative is a descriptive tool of the type, whereas the type itself refers to the imposed limits between theology and science. While in the previous chapters we treated three different philosophies in relation to their limit-relationship with science, this section will be briefer for two reasons. First, our aim in part I is primarily to trace the relationship between philosophy and science/ecology, and theology and science/ecology. Second, superficially theological understandings of nature do not differ greatly from philosophical ones, save that theological understandings of nature tend to think in terms of creation. This is important, so I will return to it in part IV, but not in terms of the two different types I've identified (declension and inflection) as both are to be located at the level of subsumption, which is, on my reading, common to environmental theology and thus not determinative at the second level.

Declension and Ecology

The declension type is not exclusive to theology. It is, in terms we will describe at length later, a population-thought that can take on different niches in different ecosystems (of) thought. For instance, one finds this population-thought in Heidegger's philosophical history of thought, which holds that Being was discovered by Greek philosophy but was subsequently forgotten. This forgetting of Being lies at the heart of our being enframed by technology, an enframing that is environmentally unsound.[3] Heidegger's declension narrative has been influential in environmental philosophy, but with regard to philosophy's relation to science it exists relative to the general subsumption model. Its function is to open the question of Being again. Within theology, the niche of the declension narrative is to "plant a seed," it has an apologetic niche, that prepares the reader for the theologians' message concerning the right way to live precisely because

THEOLOGY AND ECOLOGY   47

of the createdness of creation. This witnesses to theology's thinking metaphysics and ethics together, a vision-in-God, which is supported because within the order of creation, even in its postlapsarian form teeming as it is with violence, one may still see the way that one ought to live.

Of course, I am referring only to theologies that specifically engage with environmental thought, and theologies like Karl Barth's lie outside the discussion here. Despite the recent attempt by Willis Jenkins to use Barth's theology within theological environmental ethics, the radical separation of theology from the earth, grounded in the transcendence of Jesus Christ, appears somewhat nihilistic when confronting environmental problems.[4] One may say that Barth's theology represents an absolute declension between theology and science, insofar as science is always a science of this World, and especially a science concerned with the earth. Consider Barth's conclusion that the self-revelation of Christ was to humanity specifically and thus any attempt to build an ethics on the idea of a common concept of life, somewhat central to ecological thought, within man, beast, and plant is to be refused: "We must refuse to build either ethics as a whole or this particular part of ethics [regarding animals] on the view and concept of a life which embraces man, beast and plant."[5]

Strangely, given Barth's extreme antagonism toward natural theology, the declension type closest to Barth's position within environmental theology is the modified natural law theology of Michael S. Northcott in his *The Environment and Christian Ethics* (1996). Yet, because Northcott affirms a position closer to Aquinas than Barth, whom he never engages with in the text, his sense of decline is less radical than Barth's and can be located historically. Our "natural knowledge" can be used in the search for divine truth: "Natural knowledge of truth is available to us because the cosmos is a realm in which the being of God manifests itself as being-in-action."[6] Thus, for Northcott, the environmental crisis can be responded to through natural knowledge and that requires a certain engagement with the sciences. Yet, this is a transcendental choice regarding science that remains unacknowledged in Northcott. This transcendental choice of science calls on ecology to provide insights into nature and the environmental crisis in the light of the declension theory already at play in Northcott's theology.[7] Thus, Northcott selects the elements from scientific ecology that accord with his vision-in-God, that of a conservative and localist politics that is doctrinally orthodox within a broadly catholic position (comprising shared doctrinal trends between Anglicans, Roman Catholics, and the Eastern Orthodox). This vision-in-God is at odds with what Northcott calls "modernity," a constellation of practices and ideologies that brings together in a social form "the money economy and industrialism" that "proved inimical to religion."[8]

48 A NON-PHILOSOPHICAL THEORY OF NATURE

This constellation of practices was not only inimical to religion, though, but to the natural or created order (to use the theological terms for this sense of the word "nature"). Ecology, not unlike in Naess's description, can only provide evidence that something has gone badly wrong in the ecosystem, but provides nothing in itself for dealing with the problem identified. Thus, if we consider Northcott's transcendental choice and ask what he draws from ecology, it is nothing more than a litany of statistics relating to soil erosion, deforestation, climate change, species extinction, and the dwindling supply of natural resources.[9] There would appear to be a kind of paradox in Northcott's use of science, for on the one hand it can provide us with facts regarding the poor health of the biosphere, but on the other hand science is clearly located negatively within his declension narrative, connecting the "scientific method" with one of the primary causes of ecological degradation, "industrialism."[10] While a few pages later he does concede that some recent (at the time) scientific ideas about the holistic, interactive, and systematic nature of reality appear to have a certain fittingness with the holistic theological position he is advancing, he does not engage with them in any depth.[11] Then, further into the book, he again turns to science to provide support for his own advocacy of localism.[12] This remains, if not a paradox, a sign of the project's metatheological incoherence, as we have been treated to a long treatise on science's culpability in the environmental crisis, suggesting that what is needed is some kind of return to preindustrial forms of society, while at the same time employing certain scientific facts to lend authority to his argument. In no specific way, other than providing facts about the degradation of the earth, which are then attributed by Northcott to certain theological (metaphysical/ethical) positions, is the science of ecology presented as a challenge or spur to thought. As we will see in part II this isn't a particular failing of Northcott, but common to the division of labor between sciences and philosophy found at work in philosophy's thinking about itself.

One potential obstruction to any fruitful engagement with ecology, or science in general, may lie in his need to absolve his particular variant of Christianity from any culpability in the environmental crisis. Northcott's target here is the popular thesis put forward by Lynn White that the account of creation in the Christian scriptures is responsible for the rise of a technological and industrial relationship that is an exploitative relationship to the rest of nature. Northcott, to the contrary, holds that "the rise of instrumental views of nature has gone hand in hand with the demise of the traditional view of creation as the sphere of God's providential ordering, and with the gradual secularization of European civilization which began at the close of the Middle Ages and reaches its nadir in secularized modernity."[13] Northcott puts forward this thesis while at

THEOLOGY AND ECOLOGY   49

the same time arguing that the modern understanding of science arises out of Christian nominalists, rather than what he takes to be the much richer theology explicated by earlier Medieval theologians such as Aquinas and Augustine.[14] Again, we see the transcendental choice at work here. In one breath, Christianity is absolved of crimes against nature by referring to Christianity's varied form, and science is linked to deficient forms of Christian theology.

It is possible to confuse the difference in number of theologians discussed under the declension type to that of the inflection type, one environmental theologian to three in the next section, with a comment on the relative influence of each type within religious communities. The reality, though, is quite different as the relative declension model discussed earlier is found in the official teaching of the Roman Catholic Church as outlined in Pope Benedict XVI's *"In the Beginning…"*: *A Catholic Understanding of the Story of Creation and the Fall*, written when he was then Cardinal Joseph Ratzinger, and in his third encyclical, *Caritas in veritate*, outlining Roman Catholic social teaching. In the encyclical he sums up Roman Catholic social teaching regarding ecology in this way:

> The deterioration of nature is in fact closely connected to the culture that shapes human coexistence: *when "human ecology" is respected within society, environmental ecology also benefits.* Just as human virtues are interrelated, such that the weakening of one places others at risk, so the ecological system is based on respect for a plan that affects both the health of society and its good relationship with nature. In order to protect nature, it is not enough to intervene with economic incentives or deterrents; not even an apposite education is sufficient. These are important steps, but *the decisive issue is the overall moral tenor of society.*[15]

All the elements of the declension model are present here: an overall subsumption of science into the theological vision, an appeal to the real deterioration of the biosphere or nature, and finally an appeal not to science, or other seemingly rational forms of thought such as economic planning, but to the renewal of human society, what Benedict calls "overall moral tenor," brought about through a return to a premodern organization of thought. Benedict's dismissal of science as a partner in the forming of theological thought is, if not explicit throughout the rest of the encyclical, implicit in his unecological splitting of "human ecology" and "environmental ecology," advocating some kind of strange "trickle down ecologics" whereby if we simply respect nature certain environmentally harmful aspects of our dwelling on the earth will, the text suggests, be alleviated or disappear altogether. Never mind that part that says an ecologically informed response to our current untenable form of life is the encouragement of

50   A Non-Philosophical Theory of Nature

family planning, including sexual education and birth control, ideas that go explicitly against Roman Catholic social teaching.

In short, because science is taken as fallen it isn't even suitable as a conversation partner beyond its being a sign of human fallenness. As we saw with Northcott, the reader is asked to accept the varied forms that Christian theology may take without any recognition of the varied practices at work in science. And again there is no sense that science could even challenge theology, that, even if it is historically correct that it arises from nominalism, it could creatively respond to the shortcomings of nominalism. Thus, in the declension type, the subsumption of science is a complete disempowering of science to challenge theology in any way, except to respond to the environmental crisis *as theology*. That is to say, it responds as theology bonded to its own circularity.

Inflection and Ecology

On the whole the inflection model engages with a science understood in a more positive manner. If the declension model signals the decline of science outside of its subsumption in a strong theological framework, the inflection model identifies a more thoroughgoing decline in human thought. This means that the desire to absolve Christianity of responsibility in the environmental crisis is distinctly absent in the work of those I've grouped under the inflection model, but rather this decline is taken as an opportunity to bring together the best of science and the best of theology in a new theological thought. Thus, science and theology get inflected within a new theology, rather than set within a general notion of decline. So, the declension *narrative* is present in the inflection type, but it has a weak influence on the overall relationship theology takes toward science, whereas it is strongly determinate within the declension type. The form of the declension narrative varies though within the figures we will survey here. For instance, in Sallie McFague it is, as it was in Northcott, a problem of a worldview she locates in a "picture [of reality]," which "was a positivistic, dualistic, atomistic one that forced both God and human beings out of the natural world and into an increasingly narrow, inner one."[16] Also sharing in this weakened declension narrative is the work of liberation theologian Leonardo Boff, who locates the crisis in the "paradigm" of "unlimited growth," a paradigm that both state socialism and capitalism perpetuated "undermining [...] the basis of wealth, which is always the Earth and its resources" as well as "human labour."[17] Rosemary Radford Ruether, despite the feminist orientation shared with McFague, is a bit

THEOLOGY AND ECOLOGY   51

more radical in that she both recognizes the lack of any "golden age" from which decline could be located and locates the underlying problem within thought further back than McFague, and thus further entrenched in our culture, in the way dominant global cultures "have construed the idea of the male monotheistic God" and the subsequent "relation of this God to the cosmos as its Creator."[18] This construction of God and the relationship reality has with God has had the effect of "reinfor[ing] symbolically the relations of domination of men over women, masters over slaves, and (male ruling-class) humans over animals and over the earth."[19] While I identify more closely with this model, ultimately the relationship between theology, as one regional knowledge, and ecology, as one scientific regional knowledge, remains overdetermined by the identity of theology and its own circularity.

Ultimately the theologians grouped under the inflection type hold that science provides a valuable and necessary dialogue partner for theology. However, their understanding of how that dialogue works differs greatly. For instance, while McFague sees promise in some contemporary cosmological and biological theories that emphasize a cosmic life to things, she also holds that the earlier "picture of reality" also had, and continues to have, scientific supporters.[20] There is a certain incoherence to her understanding of the relationship of theology and science, for while she sees within certain strands support for "the organic model," a model she believes fits well with a theological statement that the world is the body of God, that God suffers in the making poor of the world, she has no reason for thinking this statement will remain theologically or scientifically valid. Or indeed, as could be argued, why that lack of future validity may not be necessary. She thus provides no metatheology, no criteria for why this aspect of science can be used as a resource for theology. This is symptomatically important since the book bases much of its positive project on a model of the relationship between God and World (as one Body) that she says "is [...] in keeping with the view of reality coming to us from contemporary science."[21] Why is it that the science of ecology is only used as a resource in relation to the facts it gives us about ecological degradation? Her own lack of attention given to ecological science is evident in her privileging of the idea of wilderness present in her view that as mostly urban dwellers we no longer know "wilderness as a yardstick," the idea being that we no longer know how nature really is or that we lack "a measure of how we have changed the world."[22] Wilderness is not an ecological concept, but a philosophical one, for the city is an ecosystem too, and often can be understood as a viable one at that.

McFague seems less interested in engaging specifically with ecology so much as she's interested in thinking theology in the light of a new

52   A Non-Philosophical Theory of Nature

cosmology, a characteristic that McFague shares with Ruether, as they both tend to draw on physicists who have tried to bridge the divide between religion and science like Fritjof Capra. We find here something familiar, something that we saw already in our survey of the philosophers of nature, for what attracts these theologians to the new physics is the way in which the subject/object split is broken down, where hard dualisms are rendered inoperative and harmful for scientific inquiry rather than the prerequisite. "Rather than assuming a standpoint outside of and unrelated to reality, from which 'objective' knowledge is possible, the observer is an integral part of the reality observed. This means one also cannot abstract fact from value."[23] So, it is because there is now a science available to these theologians that accords generally with its own understanding of reality, a world that can't be split up into discrete facts but is as a whole creation, which allows them to engage fruitfully with science. For a good theology to arise out of an accommodation of science, science first had to come to become good science, so to speak.

Yet, this itself bespeaks an underlying desire to continue to address philosophical and theological issues by way of philosophical and theological means. There is a kind of "shallow (theological) ecologism" at work here, to draw on a form of shallow ecologism mentioned earlier. Ecologism, again, referring to the use of ecological concepts, simplistically understood, to support an argument for a particular theological viewpoint. The term is usually referred to negatively, but the underlying assumption behind the term is that the ecological concepts used are simplistically understood and that is certainly borne out in the texts of Ruether and McFague (as well as Northcott). For instance, and perhaps being an early indicator of this shallow ecologism, is the constant reference to the Gaia hypothesis of James Lovelock without any discussion of the subsequent debates in scientific ecology.[24] Yet, more telling is that the inflection model doesn't present much more of ecology than the declension model did. For the declension model ecology could only give us facts about the degradation of the biosphere and give strong evidence that humanity is guilty of causing this degradation. In the inflection model we do get a positive understanding of ecology, namely, one connected to the holistic and relational picture of the cosmos given by the new physics and summed up by Leonardo Boff this way:

Consequently, the basic concept of nature seen from an ecological standpoint is that everything is related to everything else in all respects. A slug on the roadway is related to the most distant galaxy. A flower is related to the great explosion fifteen billion years ago. The carbon monoxide in the

THEOLOGY AND ECOLOGY   53

exhaust gases from a bus is related to our Milky Way. My own consciousness is related to elementary particles.[25]

However, this positive understanding of ecology is rather thin and is not
an exclusive insight into ecology, as the constant reference to the new
physics shows. What remains constant, though, is the sense that ecology
is generally a pessimistic science, a science to be used in the same way
prosecutors use forensic science in their prosecution of accused offenders.
While throughout the texts there are constant references to the culpability
of human beings in the environmental crisis nowhere is this prosecution
clearer than in Ruether's description of the historical development of ecology: "Thus ecology, in the expanded sense of a combined socioeconomic
and biological science, emerged in the last several decades to examine how
human misuse of 'nature' is causing pollution of soils, water, and air, and
the destruction of plant and animal communities, thereby threatening the
base of life upon which the human species itself depends."[26] This suggests
a very shallow understanding of the history of ecology and, of the theologians addressed in this chapter, only Boff even takes the time to outline the
specific history of ecology.[27]

Now, of course, ecology has provided valuable evidence of anthropo-
genic climate change and helped to identify the harmful effects of indus-
trialization on specific ecosystems, but to reduce the whole of the science
to a science of prosecution belies an overall shallow approach to ecology. While these theologians, in distinction to Northcott and Benedict,
are willing to change their understanding of God on the basis of new
cosmologies, they don't blend with ecology in the same way.[28] There is
then a common problem of a shallow ecologism at work in both theological models. In the declension type, as represented by Northcott's
work, there appears to be no recognition of the varied practice of science,
nor is there a separation between the transcendental practice of science
and the various controversies, theories, findings, and so forth that constitute the day-to-day life of the various sciences. While those grouped
under the inflection type appear to recognize the historical positioning
of the sciences, they do not use the science of ecology to think through
their theological model for the proper relation of World/humanity and
God, focusing instead on possibilities opened up by the new physics for
a new dialogue between science and theology. While Northcott constantly contrasted the decline brought on with modernity with "orthodox Christianity," those in the inflection type feel the need for Christian
self-criticism. There is no real difference here, though, as Northcott's criticism of nominalist theology is formally the same as McFague's criticism,

54 A NON-PHILOSOPHICAL THEORY OF NATURE

to take one example, of theology's disrespect for the body in its major traditions. For Northcott, who expands the declension narrative, and McFague, who allows a place in her ecotheology still for a strong declension narrative, and even in Ruether, who rejects the idea that there is some prior good state from which a decline began, we still find an eschewing of technology (not unlike Heidegger) as well as the urban, laying much of the ecological blame for climate change on the lifestyles of urban dwellers, even though there is a good deal of evidence suggesting that we could continue to have an urban society, and that the required asceticism need not take a localist bent, but may actually require a closer, more urban living situation that involves a great deal less meat eating. This, however, is a practical argument that does not bear directly on the problem here, it being my own instance of a "transcendental choice" that requires a greater theory of the "organon of selection" that undergirds the circular rigor of the work, which I will discuss in part III. I bring up urban ecology here only to highlight that neither theologian has internalized the ecological outlook, for they are not even able to see that the city is an ecosystem, that there is no coherent or meaningful discussion of wilderness (unless one wants to go back millennia, but few are brave enough to push their declension narratives that far), of a biosphere where humanity isn't part of the ecosystem.

Boff perhaps provides the best theology within the inflection type in terms of coming closest to a unified theory of theology and ecology. This may be in part because the declension narrative is relativized, as it is in Ruether, but in Boff this serves to direct nearly all the attention toward the "new paradigm" more than any diagnosis of the problem. We can likely attribute this to the Marxist undercurrent in liberation theology, which locates the solution to the present situation in neither some premodern nor a preurban past, but rather through the construction of some radically different future. The little attention given to such a task only serves to implicate Christianity along with the other human elements responsible for the ecological crisis.[29] Yet, if we recall the quote from Boff earlier that defines ecology as the science of relationality, we see that Boff too engages rather shallowly with ecology. Though relationality is often seen as a main tenant in ecology, and certainly the science is concerned with the relations within a particular ecosystem, it is also concerned with demarcating these relations in ecologically meaningful ways. Thus, ecologically, the slug on the roadway, while being ecologically related to the particular ecosystem, is not "related to everything else in all respects." Or, to be more precise, it's relation to everything else differs importantly in intensity. A concept that could prove fruitful in a theological context especially in terms of theorizing about theology's discourse itself.

THEOLOGY AND ECOLOGY   55

Conclusion: Ecological Thought
after the Perversity of Nature

Chapter 1 began with a discussion of the perversity of nature; how nature, as a first name for the Real, is foreclosed to any singular thought. I have attempted to demonstrate that by surveying the limit-relationship between ecology and thought, philosophy and ecology, and theology and ecology. In the first instance it was necessary to show that ecology thinks. Ecology, as a science, has responded to various determinations of its practice by philosophy and theology and this is most powerfully put forward in the ecosystem concept. In the cases of philosophy and theology in relation to ecology the aim was to locate certain tendencies common to these regional knowledges that ultimately disclose the need for a deeper consideration of ecology for philosophy and theology, or in the hybrid form of philosophical theology. For, the aim of this work is not so much to provide answers to the ecological crisis, though if it were to aid in such a project that would be welcome, but to test if thought itself can become intentionally ecological in its construction of a theory of nature. It is clear that the main strands of environmental philosophy, philosophy of nature, and ecotheology have largely remained unchanged by their encounters with ecology. Be that for reasons of a shallow engagement with the actual science of ecology or a distrust of science in general that fact remains as an invariant. What is common in all these attempts to think ecologically, be it according to their own terms as an ecological ethics or an ecotheology or according to our terms as a philosophy or theology of subsumption or one of the other types, is the remainder of a particular decision built upon faith in the self-sufficiency of philosophy and/or theology. So, the question that arises after the chapters in part I is: What if these environmental philosophies and theologies are not thinking ecologically, that is, their thought is not ecological, not merely because this is not their goal or is not a goal that would even occur to them, but rather because the structure of their thought makes it such that they are unable to think ecologically?

    This is not to say that philosophical or theological problems are avoided here. This is another reason for this long survey of our three regional knowledges, for their traditional problems and the resources they use in dealing with them will become our material for a unified theory of philosophical theology and ecology. However, the philosophical and theological materials will be treated under the sign of ecology. They will be treated as particular ecosystems (of) thought that will be judged according to a certain ideational-diversity of their populations (of) thought as a sign of the health of the ecosystem.

56   A Non-Philosophical Theory of Nature

All of this will be treated in part III. This "immanental ecology" is necessary because under the current regime of thought we can only enter into an already declared and ongoing war between philosophy and theology and the various internal wars raging between philosophers and theologians. This war of opinion is endemic to philosophical and theological thought. We aim not primarily for propositions that can be assented to or not, but a more generic form of thought that can respond to ecological reality. Thus, I will not enter into discussions about the destructive essence of modernity or whether or not the new physics is actually scientifically tenable. I agree with Boff (drawing on Koyré and Prigogine) that nonscientific forms of thought need to be employed within a broad response to the environmental crisis.[30] Indeed, the ideology of scientism, as captured within the wider paradigm of the capitalist drive toward ecologically untenable growth, has resulted in a certain mystification of reason. Yet, while I agree with Boff and others about the need for a broad but unified response to the ecological crisis, I don't think philosophy and theology will themselves become an ecological thought through the same withdrawal from ideologically determined science, which is often in practice a withdrawal from science altogether. Rather, we aim to radicalize certain elements of the preceding forms of thought surveyed.

As I will argue in the proceeding chapter, we must see all thought as equivalent before the Real, dare to go further into scientific thought alongside philosophical theology in order to radicalize the subsumption model by subsuming thought into a generic matrix. Not to subsume all thought to a preconceived thought, but to conceive of a model, a kind of quasi metaphysics that Laruelle calls philo-fiction (playing on science-fiction), from which thought moves forward denuded of its presumption to divinity, or its belief to be speaking for divinity. At the same time, this radicalizes the bonded model, but instead of bonding philosophy to one science, we understand that philosophy must always be bonded to non-philosophical thoughts within the radicalized subsumption because of the perversity of nature. From theology we see that metaphysics and ethics, dualisms in general, may be unified when one begins to think from that which is foreclosed to thought, which is the highest in thought, but this too is made generic and denuded of its particular revelation within a generic framework of revelation. From this unified perspective we may begin to respond to the ecological crisis as a crisis in thought by way of an ecological thought.

Part II

The Non-Philosophical Matrix

Chapter 5

Theory of the Philosophical Decision

Toward a Peaceable Democracy (of) Thought

At the end of the preceding chapter I ventured the thesis that the environmental philosophies and theologies discussed do not think ecologically, that is, their thought is not ecological, because the structure of their thought makes it such that they are unable to think ecologically. In this chapter I will develop this idea by exploring the structural relationship that philosophy and theology have with the sciences, which includes ecology of course, and argue that it is the self-sufficient structure of philosophy and theology that is responsible for this inability. Part II will focus on the work of French thinker François Laruelle who has, for the last four decades, developed a theory he calls non-philosophy or, more recently, non-standard philosophy. In Laruelle's view there is an intractable war between philosophies and between philosophy and other regional knowledges, in particular science. The war is intractable because, by the criteria of intellectual labor, each form of thought *operates* or *works*. John Mullarkey discusses this in his own reading of Laruelle, showing how particular forms of thought that claim to be at odds with one another nevertheless all still have some level of success sufficient to allow them to believe these forms of thoughts should persist, that they are right and helpful. Yet, the respective metadiscourse, in our case the metaphilosophy or metatheology, "imply that *only one should work*—their own. Their claims to truth are mutually exclusive. [...] *And yet they still do both work.*"[1] The sense of "work" Mullarkey deploys and the sense of ecological sense of work are essentially

60   A Non-Philosophical Theory of Nature

of the same kind. This can be seen clearly in discussions on how differing ecosystems can be said to "work," that is, to "operate," without their being a single ecosystem that is the transcendent telos of the particular ecosystems. Or, to put it in clearer, less theological language, the ecologist studies and describes the desert just as much as the rainforest, identifying its ecological relations between the living, the dead, and the never-living, its biodiversity, its resilience to disturbance, an evaluation of each ecosystem according to certain generic criteria without passing judgment according to a single transcendent ecosystem. Indeed, the ecologist tells us that the desert is a viable ecosystem that works just as much as the rainforest does, even if there are extreme differences between the two in terms of human utility, which includes aesthetic appreciation, and even though there are differences in less anthropocentric terms, as in the amount of biodiversity. In fact, within the context of an ecosystem's resilience, biodiversity is not a good in and of itself. Rather, biodiversity operates within an ecosystem as "a major source of future options and a system's capacity to respond to change and disturbance in different ways."[2] We are unable to pass judgment based on the hard number or even the percentage of biodiversity in one ecosystem over the other, but must evaluate the strength of that biodiversity within the particular ecosystem it exists within. The same is true of our ecosystems (of) thought; one cannot simply discard a particular philosophy or theology on the basis of its low quotient of diversity of "species (of) thought." In this way, by evaluating and identifying the ecological status of these various theologies and philosophies, what Laruelle calls "making them equivalent," we end the war by not participating in it within this thought.[3]

Laruelle's identification of a war within philosophy may be a stumbling block for those who want to engage with his work. However, even if we bar this word "war" from our description of philosophy (and theology and ecology) there is undoubtedly still an antagonism between philosophers, theologians, and science. This is clear in each of the ecologies, philosophies, and theologies reviewed in the preceding chapter and it would seem that, aside from a few fits and starts here and there, neither side is serious about fostering peace, what Laruelle calls a democracy (of) thought and even a communism (of) thought, between their disciplines.[4] A remark from Husserl on science serves to illustrate both the general philosophical attitude toward science as well as the theological method:

> One must finally achieve the insight that no objective science, no matter how exact, explains or ever can explain anything in a serious sense. To deduce is not to explain. To predict, or to recognize the objective forms of the composition of physical or chemical bodies and to predict accordingly—all

Theory of the Philosophical Decision 61

this explains nothing but is in need of explanation. The only true way to explain is to make transcendentally understandable. Everything objective demands to be understood. Natural-scientific knowing about nature thus gives us no truly explanatory, no ultimate knowledge of nature because it does not investigate nature at all in the absolute framework through which its actual and genuine being reveals its ontic meaning; thus natural science never reaches this being thematically.[5]

What is the cause of such aggression? Non-Philosophy does not aim, in its mature formulation, at a simple overturning of the usual philosophical hierarchy that places science below fundamental inquiry. So, from the perspective of non-philosophy, we can recognize that when the sciences, ecology included, attempted to free themselves from philosophical over-determination they did so through their own antagonistic means. They made their own philosophical claims and often, despite the structures of scientific practice, set up their own sense of self-sufficiency. Yet, this violence on the part of the scientist does not excuse the aggression of the philosopher, who polemically claims that science does not think. For the philosopher science needs to be transcendentally grounded, which philosophy does through a fundamental inquiry into the very conditions for thinking. Without this fundamental inquiry science "explains nothing." A similar stance is found in theology where theologians argue that the sciences can only make sense, can only have meaning, if they relate their discussion about beings to the ultimate Being who is God. Without God, some brash theologians may claim, the ecologist cannot even recognize a tree but only see brute, dead matter.

Is there any truth to these claims? The axiom that nature is not hidden but perverse suggests that, as Husserl claims, science cannot explain nature, though that does not have to do with science's structural inability to think but with the real character of nature as disclosed by science. In other words, it is science, specifically ecology in this instance, that shows the perversity of nature and that shows that nature may be thought from, but never captured by thought. Can we not say, against Husserl, that everything subjective demands to be understood? After all, Husserl's escape from the natural attitude by way of the transcendental subjectivity is itself dependent on natural processes, on nature as such. Furthermore, it isn't at all clear that any mutation of philosophy by science destroys its ability to make things transcendentally understandable. In part IV, I will argue, against this very idea and building off the non-philosophy of Laruelle, that a unified theory of philosophical theology and ecology is just such a transcendental (or more accurately "immanental") approach to both philosophy, theology, and nature as an immanental ecology.

62   A NON-PHILOSOPHICAL THEORY OF NATURE

In chapters 5 and 6, I will provide a more straightforward explication of Laruelle's theory. Laruelle's non-philosophy attempts to provide a theory of philosophy, not a metaphilosophy as such, but a "science of philosophy" that locates its identity or invariant structure that allows the non-philosopher to use philosophy as material within a wider theoretical framework that aims to overcome problems in philosophy through unified theories of philosophy and some other material of thought. In this chapter I explain his theory of the philosophical decision, which he argues is the invariant structure or identity of philosophy as such. His theory of the philosophical decision is important as it explains how standard philosophy is predicated on a practice of self-sufficiency. This self-sufficiency explains how standard environmental philosophies and philosophies of nature operate without any strong engagement with ecological material. This opens up to the next chapter, which is centered on Laruelle's understanding of science. This understanding is very different from the standard approach to science in both Continental philosophy (both its rejection of science and more positive projects such as that of Badiou or Deleuze and Guattari) and analytic philosophy, which tends toward various forms of positivism. This discussion of the non-philosophical approach to science opens up to a more general discussion of the practice of non-philosophy and its current "matrix." This necessarily follows because the understanding of science and the practice of non-philosophy are immanent to one another.

In the last chapter of part II, I turn to a discussion of what I term "non-theology." This is my own contribution to the practice of non-philosophy and aims to explain and develop Laruelle's engagement with theological and religious material in the light of his understanding of science. The response to the theologian will have to be more axiomatic and less communicative. For what else is the theologian attempting to argue except that science must submit itself to theology; often the claim is that it must submit itself to a particular strand of theology that proclaims itself as "orthodox." Simply put, theology must be denuded of its pretense to orthodoxy and theological material must be thought from the perspective of the "generic secular," a concept I will explain at greater length in chapter 7. This mutation of theology into simple material will allow us to see how non-philosophy repeats the general structure of theological thinking and how non-philosophy may engage with theological material. Of course many theologians are open to a peaceableness between disciplines (as many of the ecotheologians discussed in the previous chapter are) and refuse to enter into the war that some theologians have declared on all forms of thinking that do not submit themselves to their orthodoxy. The argument here is simply to provide a theory of how that peaceableness is possible. Thus this refusal of the grand theological pronouncement that "only theology saves X" calls for a non-theological supplement to non-philosophy, lest it too fall into such a temptation, a temptation that exists because of its shared general thought structure. This non-theological supplement to non-philosophy, which is found in the last chapter of part II, marks a real development of non-philosophy with regard to theology in a way that Laruelle has himself not explicitly made. Ultimately this non-philosophical and non-theological practice, with all its constitutive parts, declares peace to all—the philosophers, the theologians, and the ecologists.

Theory of the Philosophical Decision

The starting questions for a reader may simply be, why Laruelle? Why non-philosophy? In other words, what is different about non-philosophy as a practice of thinking and an organization of that thinking in comparison to the general structure of philosophy and theology present in the environmental theologies and philosophies reviewed in the previous chapter? Or, at a more abstract level, what is the relationship between non-philosophy and standard philosophical practice? The answer is that Laruelle provides a more rigorous criticism of philosophy's relationship to science than is present in other philosophers I have come across. This criticism avoids being a negation of philosophy by understanding that criticism is a way to free philosophy at its most radical, which is to say at its most immanent. By freeing philosophy in this way it helps projects that use the non-philosophical model to avoid the pitfalls of standard philosophical practice and in this way allows one to step outside the arbitrary choices given in choosing between Heidegger or Russell, Deleuze or Quine, Bergson or Badiou, Levinas or Singer (using the names of these philosophers as an index of their thought). Each is relative and one is freed from the necessity of judgment and war and thus able to select and construct alongside of material that lies outside of philosophy as such.

To fully answer these questions and bear out the claims I make about the liberty given by non-philosophy we need to give some attention to Laruelle's theory and analysis of the philosophical decision. This aspect of non-philosophy has largely been the focus of his reception in Anglophone philosophy, as in the speculative nihilism of Ray Brassier. The project developed here recognizes the importance of the theory of philosophical decision, but it will not be the main focus of the project and I will not go into great detail about it, in part because the theory is well explained in

64   A NON-PHILOSOPHICAL THEORY OF NATURE

both Brassier and Mullarkey.[6] Instead, I understand the theory from the
perspective of the matrix given in the current phase of non-philosophy
(Laruelle calls this Philosophy V and I explain this periodization in more
depth in the next chapter). In this current phase Laruelle understands the
identification of the philosophical decision as what allows non-philosophy
to free thought, specifically in its philosophical forms and forms overdetermined by philosophy, from the illusions perpetuated by standard
philosophy, by locating its relative autonomy in the face of the absolute
autonomy of the Real. This allows for the production of new constructive
and speculative solutions to old philosophical problems. By locating this
structure of philosophy it locates philosophy as material and thus it frees
philosophy to enter into more fruitful relationships with other discursive
fields.

   Laruelle claims that the philosophical decision forms the invariant
structure of standard philosophical thinking and always introduces a new
transcendence, and thus new hallucinations, even in those philosophies
that claim to be of immanence. Ultimately the theory of the Philosophical
Decision claims that

   [p]hilosophy is not only a set of categories and objects, syntax and experiences, operations of decision and position: it is animated and traversed by
   a faith or belief in itself as if in an absolute reality, by an intentionality or
   reference to the real that it claims to describe and even constitute, or to
   itself as if the real.[7]

This philosophical faith in philosophy fosters a sense of philosophical
self-sufficiency, named the "principle of sufficient philosophy" by Laruelle
and meaning that "everything is philosophizable." The philosophers
believe that they can provide a unitary thought that brings together the
constituted and the unconstituted, that it can disclose the ultimate reality
of everything, and that it does so through its own categories and objects,
syntax and experiences, operations of decision and position. At issue here,
then, is the relationship of philosophical thought toward the Real and so,
to understand Laruelle's theory of the philosophical decision, I will give
some attention to these two terms and explain their function. In order
to make the function of these terms clearer I will engage with criticisms
made by the French philosopher Quentin Meillassoux of non-philosophy,
as it captures succinctly common misunderstandings of Laruelle's project,
and explain these terms by way of comparison with Meillassoux's standard
philosophical understanding of the Real and his own critique of the overall
movement of post-Kantian metaphysics.

Structure of the Philosophical Decision and the Practice of Non-Philosophy

Non-Philosophy is the practice of philosophy mutated to act within the posture of science. Posture in this sense relates to the fact that science, in its lived practices as opposed to its philosophical form, does not take a position on an object, but has a certain practical posture or stance toward it. From this posture Laruelle created early in his career (Philosophy II) a science of philosophy, treating philosophy as the object of scientific inquiry. From the practice of this science Laruelle claims to have located the essence of philosophy in what he calls the "Philosophical Decision." In his *Philosophies of Difference*, one of the main texts of Philosophy II, Laruelle develops his first theory of the philosophical decision that is then developed more rigorously in *Principles of Non-Philosophy*.[8] In sum, the philosophical decision is located, according to Laruelle, as an invariant structure organizing all philosophical endeavors. Whereas his own non-philosophy attempts to think from the vision-in-One or what he also calls "radical immanence" that conjoins what other philosophies call immanence, transcendence, and the transcendental, the philosophical decision is simply a dyad between immanence and transcendence where, as Ray Brassier explains it, "immanence features twice, its internal structure subdivided between an empirical and a transcendental function."[9] In condensed terms, philosophy breaks up immanence through positing some empirical datum separate from the transcendence of its a priori factum (otherwise understood as the fact of givenness of something apart from its empirical appearing) that must then be brought back together through some third transcendental thing (the ego, certain conceptions of immanence, experience, etc.).[10] In slightly more accessible terms Laruelle tells us in his *Dictionnaire de la non-philosophie* that "[t]he philosophical decision is an operation of transcendence that believes (in a naïve and hallucinatory way) in the possibility of a unitary discourse of the Real."[11] This philosophical decision, like many of the main concepts in non-philosophy, has an analogue in standard philosophy. For instance, Brassier traces the structure of this decision (which we will see is repeated by Meillassoux in his critique) by focusing on its analogue with the Kantian transcendental deduction. In Brassier's helpful synthetic reconstruction of the formal structure of the Decision, dependent largely on Laruelle's essay "The Transcendental Method," he focuses on the mixture of transcendence and immanence in philosophy, where there is an initial separation of the empirical/immanent and a priori/transcendent, of datum and faktum, that is then "gathered together" and united again under some absolute transcendental authority (Descartes' "I think," Kant's faculties, Husserl's ego), which, Laruelle notes, is ultimately also some reified empirical thing and the final moment of unification where the conditioned and the unconditioned are "mixed" and shown to coconstitute one another. This final moment is a second form of immanence, but one that is transcendentally represented. It is this whole process, taken to expose the transcendental conditions for being, that is coextensive with philosophy and leads to the confusion that it is thereby co-constituting of the Real.[12] It is important to note that what Brassier traces is the structure of the philosophical decision present in post-Kantian philosophy, but the philosophical decision may take on different forms from the Kantian ones discussed in his reconstruction. Thus Laruelle says of the datum and the faktum that they are "invariants and not [...] entities or essences."[13] The point being that the structure may take a different form, that the same Kantian aspects may not be found in every philosophy, but that this invariant structure, which is triadic since it is formed of a dyad that is united (not unified) and thus identified together as a third, will be that which defines philosophical practice and leads it to confuse itself for or as circumscribing any X it claims to be Real.[14]

It seems a sweeping statement, and there is little doubt that this transcendental description of philosophy has led many to frustrated rejection of Laruelle's work, but is it true? Is standard philosophy structurally incapable of stepping outside of this triadic image of thought, which always posits a transcendence disguised as immanence? The answer to that question may be impossible to give definitively within the scope of this book, but it seems plausible since Laruelle has traced it within European philosophy's most radical philosophers (Spinoza, Hegel, Marx, Nietzsche, Heidegger, Derrida, Deleuze, Michel Henry, and Badiou, among others). Others who have taken up the task of non-philosophy have also located this structure in the thought of phenomenologists such as Husserl, Levinas, and Marion; in the philosophically influenced sociology of Pierre Bourdieu; in the antiphilosophies of Blanchot and Lacan; and stretching from the ancient philosophy of Plato to contemporary standard approaches to epistemology and aesthetics.[15] It is certainly true of the general structure of environmental philosophy (though a theory of the principle of sufficient theology will be required to explain the structural inability of environmental theology, which will be undertaken later). Such a structure will take on particularities inherent to each expression of environmental philosophy, but we can trace a general structure that unites all the philosophies reviewed in the previous chapter. We hinted at it there already when we located a split between the ethical and the metaphysical. Which aspect takes the place of the immanent and which the transcendent will vary, but for the majority

THEORY OF THE PHILOSOPHICAL DECISION   67

of philosophies it will be the stuff of the metaphysical or "natural" that is taken to be the immanent object and the ethical will be the transcendent condition for thinking this stuff within a properly environmentalist practice. The decision that separates the ethical and the metaphysical is a pure amphibology, however, since what is and what should be are confused, they shift from one to the other, insofar as both are dependent on a conception of nature that unites them. Nature in this instance is both some reified thing, nature as object to be studied, and the transcendental condition for thinking it; one must first be natural in order to naturally think. In this way philosophy bars itself from engaging with science as an equal, since science has merely provided the brute data regarding nature that can only be thought under the condition of the ethical and represented as a whole philosophically. In the instance of philosophy of nature the aim is to think nature completely as the grounds of thought, to move from the object = X to the absolute itself. In both cases this structure ultimately reflects the philosophical decision itself, for the thing-in-itself is now taken to be the split object united under transcendental philosophical representation. The object is turned into a mirror that reflects back philosophy. That philosophy then takes to mean that philosophy is able to adequately reflect back the Real itself, to circumscribe it.

Because of his focus on the Real and positive engagement with science Laruelle's project has been confused in the Anglophone reception of his work with that of Quentin Meillassoux's "speculative materialism," which also aims to think a "non-correlative" form of philosophy that has access to the thing-in-itself or the Real. It is interesting then to note that Meillassoux himself proffered a critique of non-philosophy, which contains within it all the standard philosophical misunderstandings of Laruelle's non-philosophy, at an event called "Speculative Realism" held at Goldsmiths, University of London, on April 27, 2007, the transcript of which was published by the independent journal *Collapse*.[16] In the critique put forth at this event Meillassoux even comes to confuse Laruelle's project with his own, in part because he is following and responding to Ray Brassier's work *Nihil Unbound: Enlightenment and Extinction* (2007), which uses both Laruelle and Meillassoux to advance a particular nihilistic and scientistic philosophical project. In that book Brassier claims that while Meillassoux identifies more clearly and rigorously the inherent antirealism of continental post-Kantian philosophy (which he calls "correlationism"), it is Laruelle's axiomatic stance regarding the Real that grounds any noncorrelationist realist philosophy. Brassier holds this thesis because, in his view at the time of that book, Laruelle's axiomatic method of non-philosophy avoids any recourse to "intellectual intuition" while Meillassoux has to posit it in his own philosophy.[17] Meillassoux says that his project consists

68   A Non-Philosophical Theory of Nature

"in trying to understand how thought is able to access the uncorrelated, which is to say a world capable of subsisting without being given. But to say this is just to say that we must grasp how thought is able to access an absolute."[18] In this discussion of the theory of philosophical decision I am unconcerned with setting in conflict Meillassoux's speculative materialism or Brassier's transcendental nihilism against Laruelle's non-philosophy. I focus only on Meillassoux's critique and contextualize Brassier's reading with the ultimate aim of illuminating non-philosophy's identification of the Philosophical Decision as well as the method and content developed in response to that structure or the method that Laruelle uses upon the material found in the various standard philosophies.[19]

Meillassoux's own critique of post-Kantian philosophy locates a correlation between thought and Being that is unable to account for a time anterior to the conditions for thought, that is to say it is unable to think Being separate from thought. Further to this, the correlationist thus must either explicitly or implicitly, in a weak or strong form, claim that thought determines Being (not, as Laruelle claims philosophy thinks, the Real). This leads, in part, to antirealism, but not primarily. Meillassoux says, "Correlationism is not, in my definition, an anti-realism but an anti-*absolutism*."[20] Correlationism is challenged by Meillassoux's own speculative materialism, which confronts correlationism with an aporia in its inability to think a real ancestral event outside any possibility of thought and builds a metaphysics of the absolute that focuses on primary qualities as the object in itself.[21] Correlationism is also challenged by speculative idealism, which includes for him panpsychism and vitalism, which claims that the absolute is the correlation itself and is thus dependent on the structure of correlationism. Thus, Meillassoux's goal is to champion one philosophy, an anticorrelationist speculative materialism that develops a metaphysics of Being from primary qualities, over against other philosophies. His problem is then a philosophical one and completely oriented toward questions about the Being of things. This should not be confused with Laruelle's project as seen in his development of a theory of philosophical decision.

It is at this point in Meillassoux's critique that he confuses Laruelle's conception of the Real with his own conception of a thing-in-itself as primary qualities. This is a perfect example of the philosophical decision at work—the thing-in-itself is taken to be the Real and philosophy encompasses the Real, understands it and circumscribes it, by splitting it into primary qualities and secondary qualities that are united through appearance. Whereas the Real in non-philosophy is not circumscribed, non-philosophy does not limit it to primary qualities, does not exclude anything from the Real; in fact, it refuses to think the Real, instead it aims to think *from*

THEORY OF THE PHILOSOPHICAL DECISION   69

the Real as foreclosed to thought. Meillassoux confuses this practice of non-philosophy when he claims that Laruelle's "non-philosophy is supposed to think the relation of thinking with a Real which precedes philosophy," but this, Meillassoux says, exposes a contradiction in Laruelle, for "the name 'non-philosophy' can only be constructed from the name 'philosophy' together with a negation. Philosophy precedes non-philosophy in nomination, as in the acts of thinking."[22] In other words, Meillassoux is claiming that the Real, which is supposed to be autonomous and prior to philosophy in his reading of Laruelle, is only thought in non-philosophy after the negation of philosophy. This is a misunderstanding of Laruelle's project and brings to the fore the major differences between the standard philosophical project and the non-philosophical one. First, as has already been mentioned but which must be repeated to detractors of non-philosophy again and again, the "non" in non-philosophy is not a negation of philosophy, it is a mutation of philosophical practice, which takes its posture from non-Euclidean geometry (which, of course, is not a negation of geometry!).[23] Laruelle has been clear about this throughout the development of non-philosophy. Even at his most polemical he has held that non-philosophy is not a new philosophy but a new practice that uses philosophy. In *Philosophie et non-philosophie* he writes:

> Non-philosophy is not the mass negation of philosophy, its (impossible) destruction, but another use of it, the only one which is able to be defined outside of its *spontaneous belief in itself; a practice of philosophy which is no longer founded and enclosed within philosophical faith, but that establishes itself in a positive way within the limits given by placing that faith between parentheses.*[24]

The point of non-philosophy is simply not to think the Real. The first axiom of non-philosophy is that the Real is foreclosed to thought and so it cannot think the Real in any meaningful sense, in any sense that would change the Real, but it may only think from it. It does not aim at reviving a prior philosophy or constructing a realist philosophy that grounds science or that protects philosophy from embarrassment before science. Laruelle aims to make all philosophies equivalent, to take up a scientific posture toward philosophy, in order to leave the war between philosophers, using them as simple material in an autonomous exercise that is thought from the Real.

This aspect of Meillassoux's critique, in its confusion with itself, does not really touch on non-philosophy. Its weakness arises partly as confusion of the order of Laruelle's thought. The first task is to posit an axiom that states that the Real is radically autonomous to philosophy. This axiom is

70 A NON-PHILOSOPHICAL THEORY OF NATURE

arrived at by copying the posture of scientific thought. Being analogous to the phenomenological suspension of the natural attitude, it can be called a suspension of the philosophical attitude that suspends the principle of sufficient philosophy. From this axiom non-philosophy aims to think from the Real, giving this act the name of "vision-in-One." All philosophers ask what the philosophical act consists in, but this change of posture aims at a more radical answer to the question, for when philosophers pose this meta-philosophical question they can only pose it within philosophy itself, they are unable to ask the question from outside of the very essence of philosophy.[25] By taking the scientific posture or stance, the stance from the One (as a first name of the Real), one can render all philosophies equivalent in their posture toward the Real and accountable to the Real as dependent upon it.

By taking the scientific posture against the pretensions engendered by the Philosophical Decision, Laruelle humbles philosophy. This stance posits the radical autonomy of the Real, which is to be differentiated from something like an "absolute" autonomy. What the radical autonomy of the Real means is that the Real is unobjectionable, it always outruns the structures of the Philosophical Decision, it is unrepresentable actuality: "*[T]he real is given before all givenness* [donation]."[26] This fundamental axiom, one of the most primary of non-philosophical principles, is the site of Meillassoux's stronger criticism. While I located a weak criticism in Meillassoux's confusion of the name "non-philosophy" and the order of operations in non-philosophy's methodology, there is this other aspect of Meillassoux's criticism that seems to be more damning. The criticism is directed at the axiomatic nature of non-philosophy, which is its positing of the Real "prior to philosophy" (this is Meillassoux's language, while it is more non-philosophical to say "radically autonomous to thought" or "foreclosed to thought"). Meillassoux's criticism is direct and claims that Laruelle does not prove the existence of this Real, but merely posits it— meaning that the Real is a posited Real and thus correlational.[27]

It is true that there is no hiding this axiomatic character in non-philosophy. Throughout Laruelle's corpus there are axiomatic descriptions of the Real, taken to be nonabsolute, but rather a witness to the radical autonomy of the Real. So, when Meillassoux accuses Laruelle of "merely positing" the Real he is calling on the non-philosopher to account for the very practice that non-philosophy prizes, which is precisely not to think philosophically *about* the Real and instead to think from a realist suspension that develops the consequences for thought (what is called "philo-fiction" as a mutation of metaphysics) of the primary axiom, the Real is foreclosed to thought. This non-philosophical response can only be frustrating to the philosopher of the absolute whose aim is get to the Real

THEORY OF THE PHILOSOPHICAL DECISION

= X beyond the X that is thought. While Meillassoux attempts to argue with the correlationist on philosophical grounds, the scientific posture of non-philosophy completely ignores the arguments of philosophy on the basis of a realist suspension. The question then isn't about "argument," the rules of which appear to be known only to the particular philosopher or school of philosophers, it becomes one of practice as organized by the principles of non-philosophy.

What I am terming here the "realist suspension" is, then, the response to the Meillassoux's challenge to the axiomatic character of non-philosophy. That is, the Real is pragmatically asserted through a variety of axioms, rather than circumscribed and represented as in Meillassoux's philosophy that claims all primary qualities, qualities of the absolute thing-in-itself, can be known sufficiently though mathematics. For, while Meillassoux runs aground in his terms on the necessary correlation of mathematics and human thought the realist suspension is a pragmatic style of thought that asserts the ultimately Real identity of all things. There are two aspects of the realist suspension: that all things are in-the-last-instance Real, meaning that they remain what they are, and at the same time as the Real is itself foreclosed to the thought. So, real objects may be described and known while the Real is always the deductively known cause of these real objects, not ontologically, but as One. Brassier helpfully summarizes the six corollary axiomatic descriptions of the Real as found in Laruelle's *Philosophie et non-philosophie*, which further describe the basis of the realist suspension:

1. The [R]eal is phenomenon-in-itself, the phenomenon as *already*-given or given-*without*-givenness, rather than constituted as given via the transcendental synthesis of empirical and a priori, given and givenness.

2. The [R]eal is the phenomenon as *already*-manifest or manifest-*without*-manifestation, the phenomenon-without-phenomenality, rather than the phenomenon which is posited and presupposed as manifest in accordance with the transcendental synthesis of manifest and manifestation.

3. The [R]eal is that in and through which we have been *already*-gripped rather than any originary factum or datum by which we suppose ourselves to be gripped.

4. The [R]eal is *already*-acquired prior to all cognitive or intuitive acquisition, rather than that which is merely posited and presupposed as acquired through the a priori forms of cognition and intuition.

5. The [R]eal is *already*-inherent prior to all the substantialist forcings of inherence, conditioning all those supposedly inherent models of identity, be they analytic, synthetic, or differential.

6. The [R]eal is *already*-undivided rather than the transcendent unity which is posited and presupposed as undivided and deployed in order to effect the transcendental synthesis of the empirical and the metaphysical.[28]

Some have suggested that using the One to discuss the Real as prior to Being and Alterity and foreclosed to both suggests that non-philosophy is a "negative theology" of the Real. In reality non-philosophy isn't a negation at all, but rather takes the same pragmatic posture that science does with regard to its own practices. It would be a mistake to call this a naivety, as Husserl does, for rather than a negation it is actually a more *lived* form of thinking. More adequate to the ways in which one actually practices thinking in the flow of ordinary time. That is, it takes the Real as the necessary "superstructure" for thought, rather than as that whose being or nonbeing negates the possibility of thinking in general.[29] In other words, the realist suspension, which can be found in science, is a relationship with the Real that thinks without any recourse to a transcendent self-founding.[30] Non-Philosophy is a practice, not an account of foundation, and in that way it does mirror theology and, while escaping the critiques of Meillassoux, still requires a non-theology to supplement and complete its practice by explaining the immanent practices and resistance to illusion at work in the realist suspension of non-philosophy. But before turning to that non-theology we must first explain the practice and principles of non-philosophy that non-theology necessarily builds from.

Chapter 6

The Practice and Principles of Non-Philosophy

In order to fully understand the practice of non-philosophy we need to examine the different forms it has taken through its development. These are called waves by Laruelle and by looking at the form each has taken with regard to the status of the fundamental axioms of non-philosophy we will begin to understand how non-philosophy is practiced alongside of principles rather than a law-bound method. I will then turn to a discussion of how Laruelle provides a model or philo-fiction, what standard philosophy may call both a metaphysics and a metaphilosophy, for understanding how philosophy and science may come together in a unified theory. While this may seem at first glance unrelated to the project undertaken in this work, a unified theory of philosophical theology and ecology, it is actually important as the work undertaken here benefits from the experience and mature formulation of the relationship between philosophy and science. Non-Philosophy provides the philo-fiction that allows us to treat these discursive fields as simple material, as an occasion for thought that is autonomous but foreclosed from the Real and this realization is liberating for thought as it breaks the transcendental hallucinations of standard philosophical practice, whether that practice goes under the name of scientism or vitalism.

The Axiomatic Practice of Non-Philosophy

The question of the relationship between science and philosophy was central in Laruelle's shifting from a typical philosophical practice (what he

74   A NON-PHILOSOPHICAL THEORY OF NATURE

calls placing one's thought under the "principle of sufficient philosophy") to the founding of the non-philosophical project. This attempt to think science and philosophy together is not hidden in his work and is given a central place in his own history of non-philosophy outlined in his important *Principles of Non-Philosophy*. Here he explains the periodization he has given his own work and that is to be found in nearly every book of his; on the page typically reserved for "Books by the Same Author" you will find a division of his twenty-one books into a categories labeled "Philosophy I" to "Philosophy V." Recently he claimed that he had "finally understood the principle of this endless classification, they are not movements or stages, perhaps they are phases, but most certainly they are waves, nothing other than waves, it is always the same form with slightly different water each time."[1] Laruelle's self-assessment of his work refers to how the general structure of non-philosophy has remained more or less constant throughout the 25 years since the practice of non-philosophy proper began. Laruelle calls this structure philo-fiction in order to differentiate this theory from the standard philosophical name of metaphysics, which is its closest equivalent in philosophical practice since both discuss, in very different ways, the structure or relationship of thought and being.

The general shape of these different waves, which will be discussed in more detail here, can be succinctly summed up in a paragraph. The general aim of non-philosophy is to think a transcendental realism that fosters a certain equality among objects and discursive materials. Non-Philosophy posits a Real that is foreclosed to philosophical thought, meaning that philosophy does not have any effect on the Real, the Real is radically autonomous from thought, but thought, as ultimately Real in-the-last-instance, also contains a certain relative autonomy and so any particular expression of thought can be taken as a simple material for an occasional theory. Occasional in this sense refers to the lack of self-sufficiency in those theories, a positive lack that protects against philosophical illusions of co-relation with the Real. Laruelle's early work, which includes the books listed under the heading of Philosophy I, was undertaken in large part as a standard philosopher writing works on the history of philosophy, political philosophy, and deconstruction. Philosophy II marks his break with what he calls the "principle of sufficient philosophy" after he comes to recognize an invariant structure to philosophy that limits thought by taking philosophy as, in some sense, unlimited. In this period Laruelle aims to break this sufficiency by way of a confrontation between philosophy and science, and during this period there is a simple reversal and overturning of the dominant hierarchy between philosophy and science as identified by Laruelle. In the period under the heading of Philosophy III Laruelle moves beyond just a mere reversal of this hierarchy, making

NON-PHILOSOPHY—PRACTICE AND PRINCIPLES

actual the declaration of peace to the philosopher that had been hinted at in the previous period by way of his conception of a unified theory of philosophy and science. This theory was unified, rather than a unity, because of the general philo-fiction structure that theorized both discursive practices as relative before the Real. In Philosophy IV Laruelle deploys this unified theory in a number of investigations of other non philosophical material (in the sense that they lie outside of "philosophy proper," hence why I have left out the hyphen), returning to questions about politics and turning for the first time to a serious investigation of religion, thereby deepening his concepts and elements of style, like heresy, that he had lifted from various religious traditions and that were already operative in his earlier works.

In his *Principles of Non-Philosophy* he provides a short history of the development of non-philosophy in relation to the way that the axioms of non-philosophy were modified in accordance with new materials. He claims that what changes within the structure of non-philosophy, what he now describes as "the water of the wave," is the axioms at play within it. Here Laruelle locates three distinct periods of non-philosophy, Philosophy I–III, that he thinks responds to the triadic structure of philosophy itself (understood by Laruelle to find its essence in the philosophical decision).[2] In his own estimation the work of non-philosophy, where philosophy is finally taken as a simple material that one can work with, does not truly begin until Philosophy III and is not truly completed until his latest major work, *Philosophie non-standard*. Philosophy I is characterized by what may be described in a Deleuzian way, as Laruelle's apprenticeship in philosophy. In his own words this period should be understood to have "placed itself under the authority of the Principle of Sufficient Philosophy."[3] He continues on to say, that at this stage he

already sought to put certain themes to work; themes that would only find their definitive form, a transformed form, in Philosophy III: the individual, its identity and its multiplicity, a transcendental and productive experience of thought, the theoretical domination of philosophy, the attempt to construct a problematic rivaling that of Marx, though mainly on Nietzschean terrain and with Nietzschean means.[4]

While the work here prefigured in an indefinite form the problems that Laruelle continued to consider well into Philosophy III, its true addition to the project of non-philosophy was the discovery of the principle of sufficient philosophy or the philosophical faith that *everything is philosophizable*. In the end this faith in the sufficiency of philosophy masks a correlation between philosophy and the appearance of the Real found in the

76   A NON-PHILOSOPHICAL THEORY OF NATURE

various regional knowledges philosophy is dependent upon (science, art, politics, psychoanalysis, etc.) and allows the philosopher to confuse the philosophy-of-X with X itself, through the operation of the philosophical decision described in the previous chapter. Philosophy comes to be confused with the Real itself, rather than seeing that the X of philosophy-of-X is actually a reflection of itself.

Philosophy II marks Laruelle's break with thinking under the conditions of this philosophical (self-)sufficiency, but that break is, he tells us, "more than a break or more than a new primary decision, it is the subordination of the non-philosophical decision to its immanent cause, the vision-in-One."[5] While we are already familiar with one of these two terms, the philosophical decision, the vision-in-One needs explication here. In order to overcome the narcissism that arises out of the hallucinatory splitting of immanence Laruelle situates the philosophical decision in its immanent cause—the vision-in-One. The vision-in-One is equivalent to the Real, meaning that when one thinks *from* (rather than *about*) the Real then one is thinking from the vision-in-One as radical immanence. Laruelle appears to be intentionally obscure about what the One *is*, describing it through axioms alone, because non-philosophy aims to renounce the philosophical desire-for-the-One or the thought-of-the-One that always subordinates the One to Being or sometimes to Alterity. This renunciation allows the One to become a first name for the Real that does not circumscribe the Real, but actually is the Real as name; thus Laruelle uses the two terms interchangeably depending on the operative occasion.[6] One can, however, come to know from-the-One when one begins to realize that all discourses persist through the vision-in-One, but do not in themselves constitute *the* discourse on the One. The One is radical immanence itself and thus the vision-in-One is immanent to the One itself:

> The vision-in-One properly speaking is neither a philosophy "of," neither is it a vision-of-the-World, which is only a philosophical *intuition* of the World: it is, very precisely, the vision of the One in the One and from there the vision of the World and of the Philosophy in the One. It is the experience which is absolutely sufficient for thought and for which it is not necessary to look for "Being" or "ontology," nor even the "forgetting of Being," in order to think in a way that is positive, radical and coherent; which has no need construct it with Being, in its aporias and infinite topologies, in order to give it a reality that it holds other than itself; of which it is not enough to think "in" it, nor even in using it if that use is not going to really manifest and describe its essence, to give justice [*rendre justice*] to its specificity, and not confuse it with those figures of Being and the World that are still well exterior. [...] The vision-in-One is the experience of thought that remains once and for all in the One without feeling the need or the pretension of

NON-PHILOSOPHY—PRACTICE AND PRINCIPLES   77

leaving it or even of having to reach it, even when it undertakes to look at the World, Being, Philosophy, and History "for themselves."[7]

Thus Philosophy II was founded on two complementary axioms: "1) The One is vision immanent in-One. 2) There is a special affinity between the vision-in-One and the phenomenal experience of 'scientific thought.'"[8]

To fully understand the vision-in-One one must also understand the function of the name One, as a name for the nonconceptual Real foreclosed to thought. The One and the Real are equivalent. The six axiomatic descriptions of the One that Brassier summarized in the previous chapter quoted earlier (though he did not follow Laruelle in capitalizing the word perhaps owing to a personal distaste for the way it looks in English) are actually referred, in the original text *Philosophie et non-philosophie* from which they are derived, as an axiomatic matrix concerning the One. The two words are sometimes written together as the "Real-One" and clearly, within non-philosophy, there is a central connection between the Real and the One. To put it simply, the Real is nonconceptual; it is always already foreclosed to thought, to any absolute circumscription, to any philosophical determination, to any ontological or ethical determination. The One, a name taken from philosophy but ultimately forgotten by it in favor of Being and Alterity, becomes a privileged name for the Real in non-philosophy because it is beyond Being and Alterity. It refuses to split and is at the root of other words that name radical immanence. Keep in mind that in French the One is *l'Un* and that this *Un* is found within other words that bear witness to the radical immanence of the lived, meaning here the "lived-without-life" or the actuality of the lived without the transcendental guarantee of life. There is something important to the syntax of non-philosophy operative within the French language here and this operation is difficult to translate into English. Life is determined by the lived, and it is only from the perspective of a transcendental understanding of life that the lived is another thing. Thus, within non-philosophy it is the indeterminate that is primary. Not the indeterminate as what is sometimes referred to pejoratively as "postmodern flux," but the indeterminate as that which escapes any determination, that remains relatively autonomous and equal before the Real. It is the indeterminate of the "*un/e*" carried in French as the indefinite article. Thus, there is a certain similarity between Deleuze's focus on *a* life [une *vie*] as the name for immanence in his last essay and Laruelle's obsession with the philo-fiction of the One. That is, Laruelle's particular conception of immanence is as radical immanence. I will explain this in more detail later, but for now it suffices to say that radical immanence refers to the practical actuality of immanence, rather than its metaphysical character as a plane, an objective, a proclamation, immanence in a "manner of thinking."[9] Immanence is thus always singular; it is always prior to any determination by transcendence; transcendence is never transcendent to immanence, but always determined in-the-last-instance by immanence.

The One then comes to be the site where non-philosophy develops its principles of style. Where it experiments with thinking from the Real, or looking at its material from the vision-in-One, and so to understand non-philosophy's One is largely to formally understand the practice of non-philosophy. To begin with, since the Real-One is foreclosed to thought it also comes to be referred to as the One-in-One. Non-Philosophy clones its transcendental organon from this One-in-One.[10] Or, in other words, it clones a (non-)One that will be used as an organon of selection when applied to its material and that will operate on the philosophical resistance to the foreclosed nature of the Real, formalized as non(-One). The dualism, as a thought, is in a unilateral causal relationship with the One where one aspect of the dualism, the one taking the place of transcendence, will correspond to a non(-One) while the other, taking the place of a relative philosophical immanence, will correspond to a (non-)One. The non(-One) indicates that the transcendent element of thought is a kind of negation, a hallucinatory aspect of thought that arises from the foreclosed nature of the Real-One. It is that aspect of thought that responds to the trauma of the foreclosing by negating the radical immanence of the One, reducing it to some hallucinatory transcendence of Being, Alterity, Difference, and so on, but this aspect is at the same time *actually transcendent* within that philosophical occasion, but only as rooted in the radical immanence of the One.[11] The (non-)One is the suspension of negation or the negative of philosophy and thus it does correspond to those conceptions of immanence, found, for example, in Henry and Deleuze, that resist in a philosophical way the philosophical negative, but they are radicalized here so that the (non-)One indicates its mutation of the radical immanence of the One. The last vestiges of philosophical transcendence have to be identified within these philosophies *of* immanence in order to create an immanental style of thought. A thought that is, in its very practice, rigorously immanent.

The shift from Philosophy II to III is subtler than the one that marks the move from Philosophy I to II. Laruelle came to regard the second axiom of Philosophy II, which stated that scientific thought had some privilege in thinking the Real via an affinity with the vision-in-One, as a reversal of the reigning post-Kantian epistemico-logical hierarchy. This reversal ultimately constituted a "ruse of philosophy" that allowed it to refuse "to 'lay down arms' before the Real."[12] Philosophy III begins with the suspension of this second axiom of Philosophy II in order to begin thinking from the

NON-PHILOSOPHY—PRACTICE AND PRINCIPLES   79

radical autonomy of the Real—not as a reversal of Philosophy II's valoriza-
tion of science, but in order to free the Real from all authority, even that of
science. Laruelle summarizes the history up to this point writing,

> If Philosophy I is intra-philosophical and if II marks the discovery of the
> non-philosophical against philosophy and to the benefit of science, III frees
> itself from the authority of science, that is in reality from every philosophi-
> cal spirit of hierarchy, and takes for its object the whole of philosophical
> sufficiency. So, it paradoxically corresponds to the philosophy's affirmation
> of the self, but "negatively" or in order finally to suspend it globally.[13]

Philosophy III is then the proper start of non-philosophy nearly freed from
the vicious circle of the philosophical decision. It has two major concepts
that arise from the axiomatic suspension of Philosophy II's second axiom:
force (of) thought and unified theory.[14] It is from these two concepts that
the positive project begins as differentiated from its negative and critical
forms found in Philosophy I and II.

The concept of force (of) thought is complex, but some understanding
can be had if one understands its more prevalent philosophical precursor
found in the Marxist conception of labor power. According to the Marxist
ontology labor power constitutes the movement of historical materialism
and labor power in itself is not reducible to a worker's functions or output.
In capitalism this labor power is alienated from the worker by his creation
of a product that is then given a value outside of the product itself as crys-
tallized in the form of money. The force (of) thought is similar in that it is
the organon or means though which the Real possesses a causality of the
One that avoids alienating itself in its material. That is because the force
(of) thought is a clone of the One, rather than its production or reproduc-
tion into some material form proper to it. In this way it is productive of
thought in a circular manner, but in such a way that it contains the essence
of the Real without adding to or subtracting anything from it.[15] What is
most important about the force (of) thought is its alien status. The force
(of) thought appears as an alien or, as Laruelle writes it, "Stranger" from
outside of the philosophical situation, that is to say from outside of the
structure determined by the philosophical decision, and in so doing provides an occasional solution to certain problems in philosophy. In short
the force (of) thought is, as Laruelle says, "the first possible experience of
thought."[16]

Finally, there is the concept of unified theory, already discussed earlier,
but to review remember that Laruelle means by this a unified theory of science and philosophy, of ethics and philosophy, of psychoanalysis and philosophy, of religion and philosophy, and so on: "The unified theory replaces

80   A NON-PHILOSOPHICAL THEORY OF NATURE

the affinity of the One and science with the unilateral equality of philosophy and science, philosophy and art, ethics, etc., with regard to the One
and introduces the 'democratic' theme within thought itself rather than as
a simple object of thought."[17] The democracy (of) thought is ultimately an
axiom and not a conclusion. One must begin as if a unified thinking of X
and philosophies were equal in the sight of the One in order to attempt and
think outside the problems inherent to philosophy due to its enclosure in
the structure of the philosophical decision. By treating thought as if it were
democratic, rather than a thought of democracy, one begins to truly think
from the One, as the One is itself outside of any unitary discourse and is
instead the universal discourse found in regional discourses.

     With the publication in 2002 of *Le Christ futur. Une leçon d'hérèsie* (pub-
lished in English translation in 2010 as *Future Christ: A Lesson in Heresy*)
Laruelle inaugurated a new stage of non-philosophy, Philosophy IV. It is
here that Laruelle appears to have finally escaped from the self-sufficiency
of philosophy present even in Philosophy III's constant reference to pre-
cursors in metaphysical systems such as Cartesianism and Marxism. With
Philosophy IV Laruelle has begun to produce a whole host of new concepts
from the vision-in-One (keeping in mind that this is an equivalent term
to the Real and the One itself) alongside religion, ethics, and aesthetics.
While the entire project of non-philosophy is in itself interesting, and can
be highly productive of thought outside of Laruelle's corpus as witnessed
to by the work being carried out in wildly different ways under the banner
of non-philosophy, it is in Philosophy IV that the true worth for thinking from the Real becomes apparent.[18] What, though, is the change in
axiom here? Does the change from Philosophy III to Philosophy IV constitute a change in axiom in the same way that Philosophy II changed to
Philosophy III? The answer to this question may be found in the text *La
Lutte et l'Utopie à la fin des temps philosophiques* (2004), translated into
English as *Struggle and Utopia at the End Times of Philosophy* (2012), where
he again turns to the axioms that allow non-philosophy to function. Here
they take a slightly modified form. To be more precise, they are modified
in that they are now more generalized: "1. the Real is radically immanent,
2. its causality is unilaterality or Determination-in-the-last-instance, 3. the
object of that causality is the Thought-world, or more precisely, philosophy
complicated by experience."[19] As will be discussed in the final chapter, for
Laruelle the World and philosophy are one and the same—the World is the
form of philosophy. Whereas the move from Philosophy II to III was effec-
tuated by the suspension of the axiom of science's privileged relationship to
the Real, Philosophy IV begins with the intensification of attention given
to the *complication* of philosophy in experience. That is, Philosophy IV is
concerned with what might be traditionally called philosophical problems,

NON-PHILOSOPHY—PRACTICE AND PRINCIPLES 81

but because philosophy's attention, for Laruelle, is always distracted, that is, philosophical problems are always a mirror for philosophy to gaze back upon itself, here these problems are the complication of philosophy in the experience of the lived human. The refusal of separating the conditions for human thought into empirical (i.e., a living human brain with adequate material support, including nutrition and the material conditions of a society that allows for such thought) and the traditional transcendental conditions that post-Kantian philosophy has attended to. This is an axiom, ultimately, concerned with the human as radical immanence, the human as equivalent to the Real-One.

Finally, the practice of non-philosophy, Laruelle claims, is accomplished at its fullest in Philosophy V with *Philosophie non-standard*. I have already described the matrix of Philosophy V that guides contemporary non-philosophical practice, but allow me to quickly fill out this history of the axiomatic mutations at work by locating the axiomatic shift and ultimately maturation at work in Philosophy V. Philosophy V's main axiom is ultimately a corollary to the third axiom of Philosophy IV, the one that stated its material is philosophy complicated by experience. First, a mutation of the third axiom, where the "complication" is philosophy "being introduced" to some other regional knowledge (like gnosticism or science). The corollary axiom may perhaps, following this mutation, be rendered something like this, the object of the causality of determination-in-the-last-instance, the thought-World (philosophy complicated by experience), may be mutated into a different form, a generic (i.e., nonuniversal) truth or utopia. In this way the hallucinatory transcendence of the thought-World is reduced to a symptom and when known in this immanent way is productive of generic truth.[20] This thought-World, that is, "Philosophy," ultimately comes to be "marginalized" in the name of this generic truth. So standard philosophy is understood by Laruelle as "a method that is productive of 'thought' and certain effects which accompany it, but it has that fate of wanting to give a specific image of the real, like the positive sciences."[21] By mutating, in this work, standard philosophical practice with the specific scientific material of quantum physics Laruelle creates a non-standard philosophy that is "precisely not constructed 'in the margins' of the philosophical model, it is rather that model which systematically cultivates the margin or even the marginality, whereas the generic is, if we can put it this, the margin as something sort of 'turned around' [*retournée*] or rather *turned-towards* philosophy, a space of welcome and intelligibility for philosophy itself."[22] In other words, non-standard philosophy uses material from science alongside of the productive powers of philosophy to produce generic spaces that reject marginality, hierarchy, and all other forms of transcendent judgment against the human as Real.

The Posture of Science

It is obvious, from the first of Laruelle's works surveyed here to the final remarks regarding philosophy and quantum physics, that Laruelle's understanding of science is of the utmost importance for understanding non-philosophy. Thus, following this historical sketch of the development of non-philosophy we must now turn our attention again to the specific approach Laruelle has taken toward science. It is during his work on Philosophy II that Laruelle first describes the relationship of science and philosophy and argues for a different understanding than previously given in philosophy. A basic theme that runs throughout Laruelle's work is that philosophy only ever sees itself in a grounding role for science, a philosophy of science that tends to make *thinking* about *thinking philosophically*, about its own philosophical practice. Thus epistemology is philosophy's philosophy of science in those forms of philosophy that are not even explicitly about science as such, because they place themselves in a position to speak for science, to *think for it* since science does not think. Even in philosophies like Badiou's and Deleuze and Guattari's that aim toward a more fruitful relationship between the sciences and philosophy, somewhat rare in twentieth-century Continental philosophy, there is still an unequal division of labor at work. For Badiou may think that mathematics is ontology, but regards other forms of science as unthinking and even as pretenders to the name of science. He thus sets up a hierarchy within science, where some particular science, mathematics, is said to attain the status of thought but the others are not: "[P]hysics provides no bulwark against spiritualist (which is to say obscurantist) speculation, and biology—*that wild empiricism disguised as science*—even less so. *Only* in mathematics can one unequivocally maintain that if thought can formulate a problem, it can and it will solve it."[23]

We find in Deleuze's philosophy a marked difference. Whereas Badiou's philosophy is always a quadruple object, split four ways between the domains of science (mathematics), art, love, and politics, Deleuze's is an agonistic plane of immanence without these strict demarcations. Badiou understands philosophy to necessarily think Being mathematically in a way that erases Nature, thus undercutting the status of the natural sciences (this is discussed in more detail in chapter 12). But Deleuze hews a bit closer to the wild empiricism of biology, accepting it is not masquerading as a science but is one; for Deleuze that judgment is not for the philosopher to make, but he does so hew on philosophical grounds separate from the science itself. Iain Hamilton Grant has traced how this exclusive disjunction between animal and number has dominated philosophical debates

NON-PHILOSOPHY—PRACTICE AND PRINCIPLES   83

concerning the metaphysics of nature and ultimately how this distinction between pure formalism and organicism provides "the alibi [...] for the preservation of the 'ancient (Greek) division of philosophy into physics and ethics.'"[24] For Grant, following Schelling and Deleuze, what is needed is a "physics of the All." This is, ultimately, Deleuze's philosophical problem, how to think the All, both what is formal and what is organic, which is called in his philosophy the plane of immanence.[25] But ultimately this plane of immanence is given the first name of "chaos" and chaos is the common milieu that science, logic, art, and philosophy deal with each in their own way. Deleuze and Guattari define chaos as that which makes chaotic, undoing every consistency in the infinite, rather than as an inert or stationary space.[26] In other words, chaos is not simply some "thing," but an infinite process that, in some way, differs fundamentally with the way we think. Thus, unlike Badiou (as well as Badiou's great enemy, Heidegger), Deleuze allows for different forms of thought in response to this experience of chaos. In other words, for Deleuze and Guattari, the division of labor is more horizontal than it is in Badiou. So within each domain there is located some essential practice that constitutes its response to chaos. Yet, Deleuze and Guattari present this sense of essentialism in a way that is ultimately inconsistent with other, more wild aspects of their philosophy. For *What Is Philosophy* is a book that proposes to provide a kind of metaphilosophy of philosophy as well as a philosophy of science and a philosophy of art within that metaphilosophy. Yet, that theory, despite its basis in an agonistic plane of immanence, is ultimately presented as one of limits. For the book spends most of its time discussing the barriers between philosophy and science (as well as philosophy and art) and when a discussion of the interaction between the two is given there is only the suggestion of a wider theory to explain that interaction. Thus they claim that the lines of philosophy and science are "inseparable but independent, each complete in itself" going on to say if they are "inseparable it is in their respective sufficiency."[27] Here science is found to be productive of thought, but there is no user manual for that productivity. A more productive theory that would allow for a more intentional engagement with science and philosophy is lacking.

What we find in Laruelle is a very different form of thought. His thought, as has been said before, aims to break out of the circle of self-sufficiency endemic to philosophy in order to form unified theories of thought that move outside of standard philosophy. Yet that difference is to be found in the way it fills out the practices of figures such as Badiou and Deleuze and Guattari. In this way Laruelle's non-philosophy incorporates elements of the division of labor found in both Badiou and Deleuze. Laruelle valorizes science in a similar way to Badiou's raising of mathematics to ontology, but

84   A NON-PHILOSOPHICAL THEORY OF NATURE

he generalizes that in a nonhierarchical way to include a general posture of science rather than the conclusions of a single scientific practice. From Badiou then there is this deep connection with the sciences that is theorized within the very practice of his thought. While Laruelle also presents a theory of science that focuses on its productive element like Deleuze and Guattari, this theory of the production of thought is given within a wider theory that avoids falling back on a certain spontaneity alongside of appeals to some Event that we find in Deleuze and Guattari. Along with Deleuze, Laruelle thinks from immanence, but no longer as a simulacrum, for in non-philosophy immanence is the radically lived, the Real itself, and thus must be thought from. Laruelle, in short, is able to break from the weaknesses of these thinkers and push forward their more radical elements.

In *Théorie des identités* (1992), the work that marks the end of Philosophy II before the transition to Philosophy III, Laruelle performs what he claims philosophers have not done—a transcendental, "which is to say rigorously immanent," description of the essence of science, one that is, as non-philosophy, non-epistemological.[28] Laruelle marks a path unlike any other, seemingly avoiding the problems inherent in scientist philosophies (which are, in many ways, akin to Christian philosophy in their practice) and antiscientific discourses that are more common in French and German philosophy (and those in the Anglophone world who follow them). This is because Laruelle grants science its autonomy by radicalizing a certain strand of philosophy that claims "ethics as first philosophy" recognizes science as philosophy's Other. Laruelle claims that philosophy is unable to accept science as its Other and must reassert its own identity by treating science to the same structure of division found in philosophy discussed earlier, which allows it to treat everything else as philosophizable. This means that it must rend apart science, separating out a brute factual and transcendent existence (the mixed philosophical identity of science) and the practice of objectivity, which is philosophy's denial of the realist character of science.[29] In Laruelle's mutation of Levinas's ethics of the Other to the abstract level, where the demand is made of philosophy that it recognize science as its Other and acknowledge the autonomy of science, he thereby makes philosophy the hostage of the science. Philosophy must submit to science here: "Our hypothesis takes the postural or 'subjective' realism of science for a transcendental guide or rule of its immanent theory."[30] This is a certain abstract openness to the Other, one that attempts to move from the closed and circular vision of the philosophical to that of a non-philosophical *hypothesis* that opens up a *possibility* within philosophy to think alongside its Other, science. This being the case, philosophy must understand science from the position of science; its hypothesis must be one of a certain abstract empathy.

NON-PHILOSOPHY—PRACTICE AND PRINCIPLES   85

It is the scientific attitude that consists in entrusting to science itself the elucidation of its essence, to science the recognition all way through of its radicality and derive all the consequences of its autonomy: science is for itself, at least in its cause—the Identity-of-the-last-instance—, an emergent theoretical object, a "hypothesis," or an "axiom" in the "hypothetico-deductive" sense.[31]

Before turning to his non-epistemological description of science, which is undertaken from this empathetic and hypothetical realist posture, let's turn our attention first to his description of philosophy's understanding of science. His critique here is ethical, it is a certain recognition that science and philosophy can't be reconciled into a synthesis, but that philosophy's crimes against science be recognized and understood. Science becomes both the internal enemy of philosophy, threatening meaning, and that which philosophy strives for as a seemingly apodictic knowledge: "Our experience of science is at the same time marked by a devalorization (as devoid of meaning and absolute truth) and overvalorziation (as factuality and efficacy) which are characterized by its philosophical and 'cultural' interpretation."[32] This image of science arises out of an alienating structure of intellectual labor, a division of intellectual labor into three parts. First, the philosopher admits that science produces knowledge as understanding. In French there are two words that can be translated into English as knowledge: *connaissance* and *savoir*. Science is said to produce knowledge of the first kind, which is, roughly speaking, a kind of "know-how" or "understanding," not knowledge of the essence of things, not true philosophical knowledge, but mere understanding. The second, *savoir*, is what philosophy aims at, the kind of knowledge that is sure and absolute. Knowledge (as understanding) is produced by science, but not philosophical knowledge, not philosophical *thought*. For, from Plato to Kant and to Heidegger, the claim is that "science does not think." Science dreams, its dream is to think even as it produces blind knowledge. The second division rips science in two, separating the multiple empirical sciences, which produce knowledge, from "an absolute, unique and self-founded science—first philosophy as ontology or logic."[33] This separation is the condition for philosophy's concept of science, "philosophy splits the concept of science after have separated understanding [*connaissance*] and thought."[34] Finally, the third division divides up the objects proper to philosophy and science: "To philosophy Being or the authentic and total real; to science not even beings, but the properties of beings or facts; the object of knowledge [*savoir*] is now that which is divided."[35]

What differentiates Laruelle from post-Kantian philosophy of science and from the general post-Kantian posture toward science, including those in the Anglophone work who deploy aspects of science as philosophical

86   A Non-Philosophical Theory of Nature

arguments and positions about reality taken to be the Real, what differentiates Laruelle from those philosophers engaging in a war over the status of science is that he accepts the autonomy of science, he accepts that it is free in its practice from philosophical conditions. He opposes this to the tripartite division discussed earlier in order to show "that every science, 'empirical' or not, is also a thought; that it is absolute *in its kind* [genre]; that it bears—at least 'in-the-last-instance'—on the Real 'itself.'"[36] To quickly summarize, in this early work Laruelle rejects the philosophical-epistemological tripartite division of labor, which results in philosophy's invariant approach to science or its separation of the transcendental and the empirical. Rather than following some variant of this usual philosophical approach to science, which may differ in terms of what is valued in this split, but still accepts some form of this split, Laruelle proposes a thought experiment where philosophy is mutated. This thought experiment is, as I have said earlier, ethical and demands that philosophy practice in the face of its Other. The essence of philosophy is to break up its object and then to take this break as constitutive of the Real, rather than localized in philosophical practice. Philosophy renders the object into a dyad, an empirico-transcendental doublet, that is always united (not unified) by some third term, whether that term be named "transcendence" or "immanence" and follow the different paths the chosen name determines. The posture of science, which underlies the various scientific practices, does not split its object, even when it breaks up an object into its constituent parts; it does so with the underlying realist thesis that the object is One-in-the-last-instance.[37] This thesis, Laruelle claims, is incomprehensible to philosophy because philosophy requires some third term, some transcendent (in its operation) unity to the object, rather than the radical immanence of the One-in-the-last-instance. Here, the two central concepts of non-philosophy appear: the One (Real-One, One-in-One, etc.) and determination-in-the-last-instance (which is the causal equivalent to the vision-in-One, and is also called unilateral duality).

Importantly for our project, and in accord with the reality of science, Laruelle especially rejects the second division within the tripartite division—the division of science into empirical and transcendental.[38] This division is the vehicle for the philosophical doublet the "empirico-transcendental," which is maximalized by philosophy in its splitting up of the Real-One in the Philosophical Decision. What Laruelle locates in science, and why he at first gives in to the temptation to simply reverse the post-Kantian hierarchy, is its nondecisional relationship to the One, such that science practices the vision-in-One rather than the philosophical splitting of the One into condition and conditioned, empirical and transcendental. Laruelle therefore claims that "[e]very science, even 'empirical' ones called such by

NON-PHILOSOPHY—PRACTICE AND PRINCIPLES   87

philosophy in order to denigrate them, are in reality also 'transcendental':
they bear upon the Real itself and, more than that, know that it bears a
relation to it there."[39] This is, of course, and as Laruelle makes clearer in
subsequent works, only a general or abstract resemblance, which he will
later call generic. More so, in the Real, there is no separation between the
empirical and the transcendental, there is no "thing-in-itself" *in* the radical
immanence of the Real. In Philosophy II, outlined earlier, Laruelle reversed
the post-Kantian hierarchy of philosophy and science, raising science to
the status of thought; against those philosophers who claim that "science
does not think" science is seen here to be a thought that moves beyond
itself, beyond the vicious circle of deciding upon itself, to the thought that
practices vision-in-One. While the mere reversal of the hierarchy is ultimately rejected, this thought experiment is what leads Laruelle to the discovery of the vision-in-One underlying science's practice, and as what can
modify philosophical practice. What the vision-in-One does is think *from*
the One rather than *about* the One. The One is foreclosed to thought, the
relationship between the One goes in one direction, it is unilateral, such
that thought has no effect on the One but is determined by the One. This
means that any thought of the One can only be described via axioms,
statements that cannot be thought directly insofar as their truth cannot
be proven by falsification, but are necessary and provisional for thinking
outward.

   It should be clear already that this isn't a scientism. Even though during this wave of non-philosophical practice (Philosophy II) Laruelle gives
philosophy over to science in an attempt to take on the same posture found
in the identity of science, it is clearly not attempting to speak for science,
to think for science, or to present science as the only human discipline
that can speak with any authority that we human beings may trust in.
Instead, it is philosophical thought made a stranger by being *in the style of*
science. This is how Laruelle exits from philosophy's intractable war of
opinion, the war over the history of philosophy, simply dropping out
of these conversations and building his own thought in defiance of the
war. The war of opinion can only be fought on common ground, which
Laruelle locates in the philosophical decision, so by thinking from outside
this common ground Laruelle refuses to engage in the war insofar as he
is philosophically incomprehensible. Yet, this incomprehensibility occurs
within philosophical language itself not unlike Spinoza or Levinas, even if
Laruelle's specific project is more helpful for the creation of a unified theory of philosophical theology and ecology. Laruelle mutates philosophy by
placing it in an ethical, scientific posture. This is clear in the second of the
central concepts of non-philosophy, "determination-in-the-last-instance,"
where it both explains and argues for the method of non-philosophical

88   A NON-PHILOSOPHICAL THEORY OF NATURE

thinking and is an instance of that thinking; it represents the circular rigor of non-philosophical thought, the refusal, even with regard to itself, to split objects. The concept is lifted from a philosophical genealogy, primarily Marxist, and then is recast "from the One" or in the style of science. It is then a chimera of science and philosophy, containing two codes in a concept that is One-in-the-last-instance.

The concept is introduced by Laruelle in Philosophy II, but develops and mutates through the various waves of non-philosophical practice, depending on the material it comes into contact with. So at times the weight is given to the "last-instance," at others to the "determination," sometimes determination is dropped and becomes "identity-in-the-last-instance" and at other times it becomes "determination-in-the-last-identity." In each case there is a specific and technical reason for the change, again related to the material Laruelle is engaging with, but in each case there remains a generic aspect to the concept that explains how non-philosophy engages unilaterally with the dualisms of philosophy. The dualisms of philosophy, the effect of the philosophical decision, are turned into materials by a unilateral operation called unilateralization and the possibilities for this operation are opened up by the concept of the last-instance. Whether this last-instance centers on determination or identity is not important, for ultimately they are the same thing for Laruelle. What is important is the generic core of the concept, that of the last-instance, which is not temporal, but real: "The really universal thought is theory *according to* the One, determined-in-the-last-instance by that identity."[40]

The generic concept can begin to be glimpsed in its Marxist origins, as explained and subsequently developed by Althusser. The original phrase, "determination in the last instance," is to be found in a letter from Engels to a J. Bloch where he is attempting to explain the central place of the economic in Marxist theory, but avoid the strict economic determinism that some had found therein.[41] Althusser argues that determination in the last instance is a powerful theoretical tool for refuting schematism and economism as it both shows how "the superstructures" are not mere phenomena of the base economic level, but have their own affectivity.[42] Superstructures are elements of human society that support the ordering of the base of society (worker relations to capital, for instance), often through obscuring the determinate effects of that base. These superstructures include important aspects of human society even in Marxist terms, such as religion, art, education, and the political system of the state, and so they need to be analyzed by Marxist means as well. By theorizing determination in the last instance Engels and Althusser after him are able to analyze these clearly important aspects of human society without forfeiting the importance of the political-economic aspect. Essentially the idea is that the

NON-PHILOSOPHY—PRACTICE AND PRINCIPLES   89

superstructures have effects, but they are akin to accidental effects and they do not have any strong effect on the core form of the political-economic realm.[43] Laruelle, calling determination-in-the-last-instance the "lost axiomatic of materialism," extracts and mutates determination in the last instance from its context in historical materialism and "transfers and radicalizes [it] in first Science or according to the One, which gives the concept its radical sense and allows its full use."[44] This mutation of the concept actually places Marxist thought within a transcendental framework as well, where its own claims to sufficiently describe reality are relativized and its orthodoxy challenged. For no longer does the determination-in-the-last-instance arise from the economic realm, but instead the claim about how thought itself, even thought concerning the economic realm, is shown to be determined-in-the-last-instance by the One.

Stated otherwise, "'In-the-last-instance' thus *identically* signifies that the cause is not alienated in the subject of its effect or in its effect itself and that on its side the effect retains a relative autonomy."[45] What this means is that human thought, arising as it does from any of its regional knowledges, has relative autonomy, but it has no strong effect on the Real-One itself. The strong causality goes in one direction, such that the object is One-in-the-last-instance and the thought cannot split the Real-One in itself, it is a unilateral duality. What is interesting, then, about determination-in-the-last-instance, for Laruelle, is that it is a double causality made up of two terms, which Laruelle says are

> [h]eterogeneous but the one to the other is also necessary in order to form a single [*unique*] causality, these are—or these have—that paradoxical nature of being exactly a single causality but formed by two bits that remain two and do not re-form a synthesis or a system. And yet it is not about the "difference," or the co-extension of the One and the Two, of the One which is Two and the Two which is One in a reversible way. Rather, determination-in-the-last-instance must appear to be irreversible, the One is only One, even with the Two, and the Two forms a Two with the One only from its point of view as Two.[46]

What is important about determination-in-the-last-instance from the perspective of a unified theory of philosophical theology and ecology, or generally from that of the humanities and science, is the truly pluralistic account of the material of science and philosophy. In Laruelle's terms these truly different identities, which are not unified in a third difference but remain identities, remain Two but as each One—a duality rather than a dualism. There is then no crass reductionism at work in non-philosophy. Neither the crass reductionism of scientism, which reduces things to "just X," in the case of nature "just matter," "just genes," "just chance," and so on, or the crass reductionism of theologism, which reduces nature to "just creation" (and the double sense of "just" is intended here). Since a critique of scientism, and other forms of reductionary "naturalism," is ubiquitous to environmental ethics, philosophical and theological, I do not devote attention to it here in the discussion of Laruelle but it will become important again in part IV, while in the next chapter I will discuss the relationship between theology and non-philosophy in order to short-circuit the theological negation of nature by way of the name "creation." This is necessary to both disclose the principle of sufficient theology and, in the last chapter, to free the name "creation" to become a form of material that may be used.

The Generic Matrix

First, however, we need to return again to the non-philosophical identity of science. Again we turn to Philosophy V, the most recent wave of non-philosophy, which is the most daring attempt to bring together the material of a particular science and immanental thought. As such it bears particular importance for our project, as we aim to do with scientific ecology what Laruelle has done with quantum physics. Laruelle has developed this wave in his short treatise *Introduction aux sciences génériques*, two public lectures he gave in the United Kingdom at the Universities of Warwick and Nottingham in March of 2010, and his major work *Philosophie non-standard*, published in October of 2010. In these works Laruelle has experimented with a new name for his theory, non-standard philosophy, adapted from non-standard physics. Throughout Laruelle's corpus and in almost all the secondary literature on Laruelle, both Francophone and Anglophone, the authors clarify that non-philosophy should not be confused with an anti-philosophy or a negation of philosophy. Just as Laruelle is now adapting a name from scientific discourse (non-standard physics), non-philosophy took its inspiration from non-Euclidean geometry, but it is hoped that this change in name will mark a new phase where the ideational pacifism of Laruelle's theory is expressed without confusion. For the purpose of non-philosophy or non-standard philosophy is not the destruction of philosophy, but its mutation and the solving of true philosophical problems by the means of other discursive fields that it comes together with as a unified theory.

In one sense Philosophy V seems like a nonidentical repetition of Philosophy III as it turns away from the complicating experiences of

NON-PHILOSOPHY—PRACTICE AND PRINCIPLES 91

messianism, art, mysticism, politics, and the like, and again turns its attention intensely toward science. Philo-fiction, the quantum, and the generic, the three terms found in the subtitle of *Philosophie non-standard*, are the important aspects of the matrix that characterizes this new phase and attempt to address both philosophical problems and problems within non-philosophy itself. Within Philosophy III's attempt to render certain "unified theories" there is a question of the particularity of the theoretical dyad chosen. One side of the dyad is always philosophy, for Laruelle, but even here the question "which philosophy" will immediately be asked. That is easily answered in the course of the non-philosophical construction, as non-philosophy takes all philosophies as, in some sense, true (or true-without-truth in Laruelle's formulation, meaning that the philosophy is true without taking the transcendental position of truth). Mullarkey, again, describes this aptly:

> Laruelle's project can best be summed up as a thought-experiment in the fullest meaning of this phrase—the experience of thought and the thought (of) experience—the experiment being concerned with what philosophy would become were it not representational *at all*, but rather the thing itself. By this I don't mean to take philosophy as an aspect of Mind that is the Real (even if its most "complete" aspect), for that would be just one more idealism, one more philosophical positing. Rather, the question is: what would we find if *all* philosophies, *in their plurality*, were real (and so *not* in accordance with their mutual exclusivity, their exclusive claims on truth and reality)? [...] As the most rigorous thought of immanence possible, non-philosophy allows every philosophy its truth and reality, not in the name of an epistemological relativism (more Continental philosophy), but through a hypothetical Real-ism (a kind of post-Continental naturalism).[47]

So, the claims of the particular philosophy are unimportant for the non-philosophical operation; it merely provides *philosophical material* for non-philosophy. The particular philosophy is taken *as if* any material whatsoever and as any material it is universal or "generic." The same is true of the second aspect of the dyad, be it religion or art or science. However, the question of this relationship between the particularity and universality of the chosen material, especially since this material is then changed as unified theory, had remained undertheorized until recently. Thus, a new concept brought in to support this axiom has been that of the generic. The concept has a mixed history, taking on different forms in philosophy, where it is split between a kind of philosophical anthropology and first philosophy, and in sociology.[48] The philosophical split is between the Feuerbachian conception of "generic humanity," taken up by Marxist thought, and the extension of this by Alain Badiou to the ontological, where "in the algebraic model of knowledge, the generic is the acquisition of a supplement of universal properties (those of demonstration and manifestation) by a subtraction and indetermination, a formalization of data."[49] The second source of the generic is to be found, Laruelle says, in societal epistemology, and is related to what in English is called the general equivalent.[50] This history is then subsumed in its own operation, rendering the historical conception of the generic as a constant within thought that has three distinct traits, which I will shorten here. First, "being-generic tends to present itself as a stranger."[51] In other words, the being-generic of a science or a philosophy carries with it the same "stranger" or "alien" or "foreign" quality described throughout his work as the first possible experience of thought (and treated already earlier). Second, "being-generic posseses an a priori function without being a philosophical a priori."[52] The discussion here focuses on the difference between a philosophical a priori, which grounds knowledge of the philosophical "over-Whole," and the notion of an object that remains unsplit. An object, Laruelle claims, is both particular and has universal significance as One (again explained in detail earlier). Finally, the third trait of the generic is that it "represents the chance for a duality without synthesis, it is the outline or the matrix of every duality as such, of the Two that structures science or its subject."[53] The generic, Laruelle claims, holds the particular and the universal in a duality without synthesis (a claim that is important to understand what science does). This duality, for Laruelle, is always unilateral; it is always "Determined-in-the-last-instance" by the One. Laruelle incarnates the generic as an individual person, one that is not solitary, but universal, writing, "The generic is the individual who has accepted her being universal but limited, who has accepted not being the tip or expression of the absolute, and thus resists a priori its influence."[54]

Laruelle succinctly described the generic in his lecture "From the First to the Second Non-Philosophy," given at the University of Warwick on March 3, 2010. There he stated, "'Generic' signifies that science and philosophy are no longer anything more than means or predicates that have lost their disciplinary sufficiency and autonomy; bodies of knowledge forced to abandon their specific finality in order to take up another that is generic, a form of universality that traverses their traditional domains of objects as modalities of the philosophical All."[55] Let's explicate this sentence beginning from its end and work our way backward to the middle and then explain the first half before finally bringing them together. The philosophical All or Whole [Tout] is simply another name for the vicious circle formed by the philosophical decision. Its other name is "World." The "traditional domains of objects as modalities" refers to the usual way that philosophy transforms the material of other disciplines, for example, Alain

NON-PHILOSOPHY—PRACTICE AND PRINCIPLES   93

Badiou's subtraction of a particular form of set-theory so that it may be recast as the science of the Being of being (ontology). Let's now look at the first part of the sentence, which is clear without explication, stating that rendering philosophy and science "generic" takes them outside of their own disciplinary domains, removing their disciplinary autonomy and purpose and instead of cross-breeding them *superpositions them*. In *Introduction aux sciences génériques* Laruelle critiques attempts to reclaim the word *métissage*, which has been used in racist ideology in France (related to the more familiar Spanish *mestizo*, common in the Americas), but also has been valued to create a generic practice of thought, a generic form of thought. For Laruelle this concept is ultimately superficial and falsely egalitarian.[56] But by bringing the two together in a truly generic form, separate from colonial-racist ideology rooted in quasi-essentialist forms of identity, we can now see that Laruelle is claiming that subtracting philosophy and science from their self-proclaimed domain will allow us to transform the usual philosophical use of subtraction. What Laruelle has done, finally, is found a way to remain true to the transcendental character of non-philosophy ("which is to say rigorously immanent").[57] Yet he is able to turn away from drawing on particular philosophies and instead draw on particular aspects of a regional knowledge. Laruelle had already been engaged in this practice in his works on religion, as he turned to particular experience of religion (Christianity, particularly the legacy of messianism in the West, mysticism, and gnosticism), but now he has turned to a particular science.

Laruelle describes the current wave of non-philosophy in this way: "So let this be the formula of non-philosophy renewed or renamed as generic science or non-standard Philosophy: *the fusion of science and philosophy under science, fusion under-determined in-the-last-instance by science, specifically quantum physics. This is our guiding formula, that which we call the generic matrix.*"[58] Note the subtle difference here between the "Determination-in-the-last-instance" by the Real-One that is common to both philosophy and science and the "under-determination in-the-last-instance" of philosophy by science. This remains an egalitarian relationship, insofar as the Real is concerned, but philosophy is said to "undergo" science or "goes under" science in the same way that one "goes under" anesthesia. The relationship between the person and the anesthesia is not a hierarchical one in any strong sense, and in the same way the immanent field fashioned in this generic science is a kind of idempotence.[59] Our project, as developed in this work, of an immanental ecology builds off of the shape of Philosophy V's wave; where Laruelle has submitted philosophy to the conditions of a particular form of science, quantum physics, we do the same with philosophical theology and ecology. Thus, in the same way that Laruelle speaks of certain *real* quantum aspects within philosophy, going so far as to say

94   A Non-Philosophical Theory of Nature

that "the science of philosophy [i.e., non-philosophy] is a *quasi* quantum physics of concepts" where "the generic matrix is an experimental chamber that allows for a struggle or collision of physical and philosophical particles in order to produce new knowledge," we will speak of philosophy and ecology where the generic matrix is a *real* ecosystem (of) thought.[60] Like Laruelle we aim for this to move beyond mere metaphor and to actually produce new knowledge, or, rather, a new way of thinking ecologically that will add to knowledge a unified metaphysical ethics adequate to the findings of ecology.

Chapter 7

Non-Theological Supplement

Before moving on to our construction of a unified theory of philosophical theology and ecology we must deal more directly with Meillassoux's strong criticism of non-philosophy. Remember that this criticism was directed at the threat of dogmatism in non-philosophy, specifically with regard to its claim to think from the Real. But the reason we have to deal with this strong criticism is not due to philosophical rules or on philosophical grounds. Rather this general axiomatic structure of non-philosophy, where the Real-One is foreclosed to thought and yet thought has relative autonomy before it as thought *from* the Real-One, is formally similar to theology. This is especially true of those theologies, like environmental theology, that aim to understand the relationship science and philosophy have toward reality. If this is true, and I show its validity later, then there is a question of why I am engaging in the non-philosophical project rather than simply a theological one. To answer that question I must first address Laruelle's own work on religion, which has been the dominant material of Philosophy IV and the source of a number of central concepts such as heresy. But additionally I must also provide a theory of theology that locates its own particular self-sufficient character.

The Non-Philosophical Identity of Religion

As already discussed, Laruelle's "science of philosophy" is often misunderstood by philosophers, even those friendly to Laruelle's vision, to be an outright assault on philosophy; non-philosophy is confused with

96   A NON-PHILOSOPHICAL THEORY OF NATURE

antiphilosophy. Alain Badiou, upon being asked about Laruelle's work in
an interview, remarked:

> I have difficulty in understanding Laruelle [laughs] especially regard-
> ing the question of the Real. The strength of philosophy is its decisions
> in regards to the Real. In a sense Laruelle is too much like Heidegger, in
> critiquing a kind great forgetting, of what is lost in the grasp of decision,
> what Heidegger called thinking. Beyond this, and not to judge a thinker by
> his earliest work, his recent work has a religious dimension. When you say
> something is purely in the historical existence of philosophy the proposition
> is a failure. It becomes religious. There is a logical constraint when you say
> we must go beyond philosophy. This is why, in the end, Heidegger said only
> a God can save us.[1]

Let us set aside the incoherence of some of Badiou's remarks concerning
religion, especially considering his own interactions with orthodox insti-
tutionalized Christianity, and let us also set aside the usually frustrating
(for philosophers) response of non-philosophers that of course Badiou, as
a philosopher, has difficulty understanding Laruelle's non-philosophy.[2]
Taken outside of its polemical context this philosophical dismissal of
non-philosophy remains instructive insofar as it identifies a religious
dimension at work in non-philosophy. Badiou isn't merely addressing
Laruelle's work on religion, but making a claim that there is a religious
aspect underlying the practice of non-philosophy itself. I think he is right,
though I have difficulty understanding Badiou's critique, and so think he
is right by accident and that the consequences of this religious character
(what might more accurately be called a theological character) are radically
different than what Badiou suggests.

Laruelle does not build a wall of absolute demarcation between philosophy and religious thought, but neither is he simply accommodating to
religion as it presents itself. While he does not hold back when critiquing
philosophers, even those like Deleuze and Levinas for whom he nevertheless expresses a great deal of respect, Laruelle equally does not let religion
go by without an equal amount of vitriol. For if philosophy operates with
a kind of theoreticism that denigrates Man, religions are the site of terror,
providing all sorts of hallucinatory justifications for the murder of Man
from an imagined transcendent source. Laruelle's understanding of religion thus shares much with those who think of religion as a construction
of forces, like Talal Asad and Deleuze and Guattari. Religion is a construction of forces, yes, and these forces ultimately point to a real human cause
of religious practices, but religion cannot be understood via a universal
category captured by an anthropological description, thick or otherwise.

NON-THEOLOGICAL SUPPLEMENT   97

For the operation of non-philosophy upon religion does not aim to merely describe religion any more than it aims to eliminate or protect it in the name of either a liberated philosophy or an enslaved philosophy. Instead, non-philosophy aims at appropriating religion: "Axioms and theorems, these are our methods, us men-without-philosophy, so that we can appropriate religion and adapt the divine mysteries to our humanity rather than to our understanding."[3]

Importantly, this method does not involve blending philosophy and religion together, but treats them as relatively autonomous within a duality that is ultimately in a unilateral relation to the Real-One. Laruelle must deal with religion, for religion is actual and thus Real, but also because religion has been the *occasion* and *material* of struggle against Worldly Authorities: "The paradox is that it is above all from the sides of the religious reality, in its dualysis, that the occasion itself is found for an emergence of subjects as Futures."[4] In the construction, not of the future, but of human Futures (like Moderns or Ancients), religion appears in world history as an instance where human beings, not gods, struggle in-immanence with and for the World. In his *Future Christ*, Laruelle aims to make use of the specificity of Christian religious material to first alter the practice of philosophy by introducing the experience of heresy into philosophy, and then to perform a non-philosophical operation on religion to put it to human use. He locates what is different between philosophy and religion via the same non-philosophical dualysis—that is, in terms of the relationship between their Authorities and Strangers:

> There is a difference here from philosophical systems that are partitioned according to the dominant (but not unique) axes of truth and appearance (or illusion from the point of view which has as an object the theory of that partition), for a religion has as its principle or dominant difference that of orthodoxy's division, from the rigour of orthology (as the policing of opinions or dogmas) and heresy, which it sometimes mixes with the philosophical that it anyhow cuts again.[5]

For, in the duality between Authority and Stranger, heresy becomes the organon of radical immanence determining in-the-last-instance the human identity of religion.

Thus, religions are not only sites of Authority, though this is their non(-One) aspect, but they are also, in their particularity, *occasional causes* for human struggle, struggles as Strangers. This aspect of religion is obscured insofar as philosophers of religion tend to focus on the orthodox aspects of any particular religion. Even when a philosopher aims for a radical critique of religion or a radical appropriation (as in Badiou's own

98   A Non-Philosophical Theory of Nature

work on St. Paul), they do not look to the victims of religion, that is, to the *religious* victims of religion decried as heretics by the orthodox. Instead, they play with the orthodox material, ignoring the heretical material, even if they are unconcerned themselves with falling into heresy. This is why Laruelle, in his own non-philosophical working with religious material, gives a great deal of attention to the various strains of thought collected under the name of Gnosticism. For, though Christian philosophers and theologians may identify all aspects of modern Western society with Gnosticism, there are no actual Gnostics left in the World. The "Gnostic question" prefigures the "Jewish question" and what is fast becoming in the West the "Muslim question" and shares with it the fact that, for both questions, the answer from the side of orthodoxy was found in fire. Thus, in the unilateral duality of religion, where the dualism is one between Authorities and Strangers, orthodoxy and heresy, it is always the heretic, as subject-in-struggle and not merely a passive victim, that determines-in-the-last-instance the identity of the particular religion that non-philosophy may then work with. In short, non-philosophy demands a unilateral thinking from the subject-in-struggle, rather than the victim or the orthodox.

Rather than philosophically working out problems inherent to certain religious thoughts, Laruelle gives "first names" to identify these occasions of the Real (i.e., their identity as Stranger) within a particular religion. This, then, is the meaning of that strange appellation "Future Christ": from the material of religion non-philosophy removes "the future" from its inscription in a Time-World, inscribed as it is within a philosophically determined understanding of the future. Instead Laruelle inscribes the future in the radical immanence of Man. The future is given its identity only as it is "lived without purpose," a future radically immanent (to) anyone and thus without telos as a subject formed in that radical immanence.[6] This appellation is derived from three sources that are blended within "the-Christianity," or Christianity as formed by the Authorities or orthodox, which are separated via a non-philosophical naming of them:

> The first is the properly Gnostic experience of the definition of man by the primacy of knowledge over faith, an untaught or unlearned knowledge that we must radicalise as Man-in-person, Lived-without-life or even as the Real. The second is the more general heretical aspect, of the separation with the World, here extended and universalized beyond its Christian and even Gnostic aspects. The third is the specifically Christian aspect of universal salvation, for the World and for every man, that works through the person of Christ, which we must also radicalize in a Christ-subject.[7]

NON-THEOLOGICAL SUPPLEMENT   99

In this sketch of Laruelle's working with religion and philosophy the particular power of non-philosophy is revealed. For there is certainly a wild, heretical freedom at work that offends the scholarly tone of philosophy of religion and the piety of theologians patiently working out the Truth through faith and/over reason, and this heretical freedom comes from non-philosophy's beginning from axioms derived from the Real-One, rather than from the philosophical history of Being or Alterity. But, non-philosophy's declaration of the One also restrains non-philosophical naming and provides it with a certain amount of theoretical rigor as that naming must, by the very same axioms, work through the material as actually given.

In his incredibly clear introduction to non-philosophy, Jean-Luc Rannou remarks that non-philosophy has the singular ability to respond not only to transcendental questions such as "What is religion?" but also to those singular questions such as "What is the Qur'an?"[8] Non-Philosophy has this generic ability because it aims to think equivalently as both science and philosophy, theology and philosophy, art and philosophy, erotics and philosophy and it calls these equivalencies "unified theories" that perform a real democracy (of) thought.[9] This is the task before any philosophy of religion separated from its authoritarian form, a philosophy of religion that is non-philosophical, to consider both generic religion (rather than religion subsumed into a universal category) and occasional particulars (like Christ and the Qur'an) from within the radical immanence of Man determined-in-the-last-instance by the Real.

At this stage I must step back from an exposition of Laruelle's work on religion and consider the possibility of a non-theology as both the name of a non-philosophical philosophy of religion, as presented in his work, and as a science of non-philosophy. The first is obvious enough and follows Laruelle's own limited remarks on the possibility of a non-theology. He calls non-philosophy "a human mathematics," a formulation he opposes to "Leibniz's conception of philosophy as a 'divine mathematics.'"[10] From non-philosophy springs a number of new possibilities for thought, one of which he calls "non-theological." This non-theological thought appears to be essentially what I've described earlier: a thinking of religious material under the aspect of Man in his radical immanence as minority, an "inversion of the philosophies of transcendence and of the divine call," the construction of a future against and for the World, and so on.[11] The point here is to use religious material to challenge philosophical practice and to transform the material of religion so that it is no longer a golem, but once again any material whatsoever.

Non-Theology as Theory of the Principle of Sufficient Theology and Internal Non-Philosophical Heresy

By giving the name non-theology to a non-philosophical philosophy of religion I am indicating that non-theology should begin with the same axioms as non-philosophy. However, as I said in the introduction to this chapter, it has to deal with the theological material that infects philosophy of religion as theological material and in turn creates axioms in response to them. It is here that non-theology becomes a name for the science of non-philosophy. Laruelle began the work of non-philosophy by first locating, and then taking a heretical stance toward, the principle of sufficient philosophy. This principle, Laruelle tells us, lies at the core of philosophy more so than any other philosophical principle (such as the principle of sufficient reason) and it is, in itself, not a philosophical principle at all insofar as philosophy is unable to see it. The principle of sufficient philosophy lies outside of philosophy's vision much in the same way that Narcissus does not see the pool that reflects his image back to him. It is thus only non-philosophy's refusal of this principle that brings it into vision. Again, the principle of sufficient philosophy can be summed up in the belief that everything is philosophizable. In this way philosophy gives itself a fundamental or necessary status in the discourses in which it shares (philosophy of art, political philosophy, philosophy of science, philosophy of religion, etc.) and as co-constitutive of the Real. Laruelle has said time and time again that non-philosophy does not aim to overcome or destroy philosophy. The principle of sufficient philosophy is merely identified as a fact about philosophy, which may explain its many failures and that, once identified and turned into material, may be used in other ways as well. It is, as such, simple material for future human use.

Laruelle, we have seen, attempts to use this material while thinking according to the Real, seeing through the vision-in-One according to and not about the Real. In this way, the Real appears to take on a quasi-divine character insofar as the Real can only be described via axioms. The nature of an axiom, however, is that it is fundamental for some system but cannot itself be proven directly. One must simply work out the system from the consequences of the axiom upon the material presented and its validity will be given if the system works.[12] Yet this axiomatic approach is the only way to actually refuse the philosophical decision as it makes the decision relative to the Real. This method decides nothing, rendering everything equivalent before the Real, finally escaping from the principle of sufficient

NON-THEOLOGICAL SUPPLEMENT   101

philosophy as it throws itself prostrate before the Real—non-philosophy has and recognizes its limits.[13]

There nevertheless remains a temptation to philosophize, for who can think according to the Real and not ask about the nature of the Real itself? Such is a temptation to heresy, but also to orthodox codification; that is, it is a *temptation to theology*. Laruelle's axioms become, as is suggested in his *Future Christ*, a form of unlearned knowledge [*savoir indocte*], differentiated from the learned ignorance of Nicholas of Cusa. Unlearned knowledge is not mystical obfuscation, but the unlearned knowledge of the Real that is radically immanent in Man-in-person, as Man-in-person is the "performation" of the Real, and from which one necessarily proceeds. There is then a similarity one may draw between non-philosophy's method and theology. Theology has its own self-sufficient problem analogous to philosophy's—the principle of sufficient theology. This is different from philosophy's narcissism and may find some elucidation by a comparison with the other figure in the myth of Narcissus, the nymph Echo.

The history of all theology hitherto has been that of the interplay between echo and control (the figure of Hera). Theology, it is often said, has no object proper. It is simultaneously simply in the service of the central event of faith (for Christianity the death and resurrection of Jesus Christ), claiming to merely echo that event, while also its complex task has been to codify the truth of that event into some sort of universal doctrine. The Creeds perform this function of theological determination brilliantly as a perfect instance of learned ignorance.[14] The Creeds respond to the historical heresies, and one may generalize about heresy by claiming that, in contradistinction to orthodoxy, they always say too much, either making a claim to learned knowing or to radical gnosis (Laruelle himself discerns this very difference between his unlearned knowledge and the principle of sufficient heresy).[15] At the same time, the Creeds go on to say quite a bit, all of it very learned, which is to say, with Laruelle, all very Greek and sometimes, though very rarely and in a qualified way, Jewish, but all of it quite ignorant of the radical immanence of Man. Echo and control is learned ignorance.

Non-Philosophy appears to mimic theology in its thinking from the Real and not of it. On the one hand, Laruelle has already noticed this and tried to differentiate non-philosophy from theology, specifically in its negative mode, by highlighting two important differences. First, the "non" of non-philosophy is not a negation, but is actually a positive operation within thought from the Real. When the (non-)One or "clone" of the Real, which is the form of philosophical immanence at play for non-philosophical thought, takes on the "non" it is always as a kind of superstructure to the infrastructure. It models the Real in practice, not representation. Thus, the "non" is not about ineffable statements such as "the One is Being beyond Being," but of positive practices within thought. Second, Laruelle claims, the (non-)One is an irreversible static *effect* of the Real, it is a dyad unilateralized from the perspective of the Real. The implication is that other negative forms of theological thought posit a negation that is taken to actually affect the Real itself, even if this goes under different names (Hegel and Hegelian forms of theology). The question of language is crucial here. While in negative theology language is always taken to be insufficient to describe God, within non-philosophy language is contingent on the One and can then still provide the means for an adequate description and in fact this contingency requires that it be described, but only in the last-instance.[16]

This differentiation, however, does not say enough yet. Theology is not reducible to negative theology, and the more sophisticated forms of negative theology are always combined with a positive project. There is a similarity between non-philosophy and theology that goes down to the level of axiomatic practice. To see this simply replace the Real with the name of God. Theology thinks from God and not of God (in the same way that philosophy would think of God). Theology cannot think of God without first thinking from God and in this way theology is an axiomatic practice like non-philosophy. Yet it is this very axiomatic aspect of theology's practice that underlies its principle of sufficient theology where everything is theologizable because theology's nonobject, God, is related or even meta-related to everything that is. In non-philosophy's methodological cloning of theology, how does it avoid its own self-sufficiency? The principle of sufficient theology is clearly in a different register than philosophy's self-sufficiency principle in that it does not claim to have a privileged place in the thinking of everything self-sufficiently, but as auto-donation or auto-givenness of Divine sufficiency from its own notion of God that functions, with various differences, in a structurally similar way to the Real of non-philosophy. Laruelle suggests in *Future Christ* that it is the figure of the heretic that must be taken up and that the Gnostic Christ is a model of heresy. Yet, the historical Christ reportedly wanted to draw all things unto himself, and, as we have seen, Laruelle locates this universal salvation as one of the sources of his appellation "Future Christ." Can one still have this sort of theological universal, even as cloned in non-religion, and avoid theology's principle of sufficient theology? If so, then non-philosophy needs to be unified with the practice of non-theology in order to overcome the temptation to this principle.

This practice operates along two axioms: (1) the Real is foreclosed to authority and tradition and (2) what is true(-without-truth) in theology is what is most generic and thus what is most secular. The operations

NON-THEOLOGICAL SUPPLEMENT   103

of these axioms are largely already at work in the practice of Laruelle's
non-philosophy, but they have not been developed in relation to theology proper nor in relation to non-philosophy's own practice within a
non-philosophical community that exists outside of the specific work of
Laruelle. This will be the focus of this chapter's conclusion before we turn
to the development of an immanental ecology in part III.

The Real Foreclosed to Authority and Tradition

The question of authority is at play throughout Laruelle's work, but its most
mature formulation with regard to Christianity, or "the-Christianity" in
the parlance of non-philosophy, a unitary form of thought that considers itself sufficient, is found in his *Future Christ*. Here Laruelle takes
up a common trope in his theory, that of the privileging of minori-
ties, of the individual Man-in-person, over that of authorities, or the
World. The World is, in non-philosophy, a name for the confusion of
some form of thought with the Real (hence we saw that in *Philosophie
non-standard* Laruelle calls philosophy a thought-World). During the first
phases of his career, Laruelle is concerned primarily with philosophy and
thus World often refers to the confusion of philosophy with the Real.
"Worldly thought" means "auto-sufficient thought," thought taking itself
as distinct from the Real, and thus thought that has fallen into hallucinatory error. Mullarkey aptly sums up the World when he writes that
"all philosophical thought is really about itself, it is auto-sufficient. Its
so-called world—x—is actually a mirror of itself."[17] With regard to theological forms of thought this reflective form is complicated insofar as the
reflection is always double. This double character of theological reflection relies on its tradition. Standard theology, called thus to make room
for a non-standard, heretical, or non-theology, claims to be dependent
upon a tradition. It looks to the tradition for its content, which it echoes
in its own voice. But this tradition is itself ungrounded, is itself but an
occasion, and in reality the tradition and authority are structured as an
amphibology since the tradition is but the discourse of authority and
authority derives its power from this discourse. It is another form of the
World, of a claim to self-sufficiency, of orthodoxy, which all new forms
of thought must be plunged into. This is common to all forms of Worldly
thought, where the individual, where radical immanence, is plunged into
some other aspect of the World, never known in itself, but always medi-
ated through the structure of the World, which is to say by way of the
Authorities.[18]

104   A Non-Philosophical Theory of Nature

In *Future Christ* Laruelle thinks from the perspective of the murdered Gnostics of history as a particular form of universally persecuted heresy. This is not a denial of the horrors of the Jewish Shoah and it is not in any way a justification for the many crimes committed against humanity by itself. Rather, it reveals something beneath the particularity of the name "Jew" or "Tutsi" or "Shi'a." It reveals that an individual human is murdered as a human. That the human endures crime. Laruelle puts it this way:

> The heretics reveal to us that man is in an ultimate way that being, the only one, who endures crime and is characterized by the possibility of being murdered rather than simply persecuted and taken hostage, exterminated as "man" rather than as "Jew." Why ultimate? *Because man is without-consistency, he is on principle, in contrast to other beings, able to be murdered, he is even the Murdered as first term for heretical thought and for the struggle that it performs.*[19]

I should note here, though without developing it further until the final chapter, that I do not hold to Laruelle's anthropocentric characterization of man as such and consider it a form of philosophical determination of the Earth that has remained within his theory. This is hardly a reason to give up on non-philosophy, though, since it calls for a kind of "permanent heresy," a constant pragmatic return to its own theory to deepen its practice. So, from the perspective of ecology there must be some room for a "crime against the biosphere" perpetuated, as the crime against humanity is, by some aspect of the biosphere against itself. But, this focus on the *minority* status or radical individual identity, *precisely because it is without-Essence,* of the human distinct from the forms of unitary identity that are bestowed upon human beings by the World discloses the radically foreclosed nature of the Real to authority and tradition. Instead, the Real-One is always a challenge to authority, always an "outside-memory" that is lost to the Western form of memory, but that is at the same time not lost because it is the essence of thought's non-consistency as always insufficient to think the Real.[20]

Thus, heresy is the privileged form of non-theological thinking, because it is in its immanence always inconsistent, always the shared inconsistency that marks the identity of the human. There are, of course, majoritarian or authoritarian forms of heresy and concerning these Laruelle remarks, "What is more hopeless than a *Principle of Sufficient Rebellion*," but these can be differentiated from heresy as struggle.[21] Laruelle delineates this differentiation in *Future Christ*, tracing the differences between war, or the *Agon* of philosophical absolute immanence (what we located in the previous chapter as a simulacra of immanence), and the rebellion of historical

NON-THEOLOGICAL SUPPLEMENT 105

Gnostics. In the case of war and rebellion it is always a matter of an under-lying authoritarian logic, a "because of." The rebellion of Gnostics against Christian philosophy is always "a reaction of auto-protection against aggression."[22] While the non-theological point is always to raise as primary that which is not autoprotective, that which is an "(immanent) because," of that "revolt that commences and does not cease to commence in each instant, proletariat or not, exploitation or not. But if it has in itself suf-ficient reasons to start, it has only too many of them and cannot make a cause of them."[23] In other words, struggle, when separated from even rebellion as a minoritarian form of authority, is separated from the World in general. It is a generic practice.

The Generic Secular as True-without-Truth

The structure that is shared by theology and non-philosophy differs on the fundamental level of practice. If non-philosophy aims at a generic practice, one that is *from* the One as rigorously immanent in-One, theology always aims at a universal practice that is rooted in something transcendent to its language. Theology, in all its monotheistic forms, and despite whatever mixture of negative and positive statements it produces, must always sub-sume the individual into the absolute and, furthermore, into the form of the human proclaimed by the authority of tradition. It must do so because it is God, the highest that can be thought, that provides the cause for the tradition. Each tradition begins as the struggle with a particular experience of God that becomes static and taken to be sufficient for that struggle.[24] Of course, this is not how any particular religion would present itself, for the invariant apophatic character of monotheistic religion allows theolo-gians to claim that the tradition isn't sufficient to circumscribe the being of God while that tradition-structure still remains the only sufficient way to think about God in a way that avoids idolatry. Or it may even allow them to claim, as is the case to some extent with the Protestant theolo-gian Karl Barth, that this tradition fosters its own dissolution, its own form of non-religion, which is ultimately the true faithfulness to the God witnessed to by the tradition slowly withering away in the harsh light of revelation. The non-theological supplement to non-philosophy retains the generic practice of non-philosophy, its general philo-fiction that encour-ages a kind of wild hyperspeculation, and so it must be, in some sense, "secular." Laruelle himself affirms this writing, "Non-Philosophy affirms a 'secularity' [*laïcité*] by principle, as universal as it or philosophy can be. It refuses conspicuous religious adherences in thought, and even the ruse of non-religion, which non-philosophy has the means to unmask and analyze."[25] The meaning of this secularity, which I have also called the "generic secular" and, elsewhere, "secularity-without-secularism," will be explored in preparation for the final chapter and before moving on to the next.

According to the conservative Anglo-Catholic theologian John Milbank "[s]ecular discourse [...] is actually *constituted* in its secularity by 'heresy' in relation to orthodox Christianity."[26] This claim is supposed to call the secular philosopher back to the fold of the Christian tradition, where their ideas find true fulfillment. However, this heretical aspect of modern philosophy has not actually gone far enough. Every instance of representation, where the Real-One is said to be finally thought adequately, which is in actuality every form of authoritarian thought, *calls forth* heresy as the actual human-in-person struggles with that authorities' attempt to alienate their actuality within a auto-positional transcendental field. Thus, there was a particular secular, one that claimed to be universal and absolute, but was in reality a post-Christian secular, a form of the secular that presented itself as an authority before which human beings, particularly non-European human beings, had to prostrate themselves. This is the form of the secular, which I will refer to as simply "secularism," that is rightly resisted by contemporary postsecular discourses. This resistance has not simply been intellectual, but also political, since this imperial secularism ultimately combined both. Talal Asad aptly sums up the nature of this secular writing: "Secularism is not simply an intellectual answer to a question about enduring social peace and toleration. It is an enactment by which a *political medium* (representation of citizenship) redefines and transcends particular and differentiating practices of the self that are articulated through class, gender, and religion."[27] This form of secularism was indeed parasitic on Christian orthodoxy, though that managed to still pit a form of Christian identity against other forms of identity in the colonial expansion of Europe. Secularism is thus a negative movement of thought. It aims to negate real human beings in their identity-practices because those practices, if allowed to proliferate, would constitute a rebellion against the dominant World or Authority. This is not mere academic speculation here, but is happening yet again in Europe without shame as in Merkel's Germany where she has proclaimed, in the light of the relationship between a minority Islamic culture and the dominant secular Germany culture, that multiculturalism has failed and that *new immigrants must adhere to Leitkultur.* Or in Cameron's Britain moving toward more draconian immigration policy as the country begins to reflect a "muscular liberalism." Or in Berlusconi's Italy (and it remains Berlusconi's at the level of culture) where Roma are required to carry identity cards with them and are said to be the main threat of raping "Italy's women." Or in Sarkozy's and Hollande's France, which has

NON-THEOLOGICAL SUPPLEMENT 107

continued its assault on non-Christian traditions, culminating in the ban on the wearing of any face covering in order to outlaw the burqa and niqab worn by a small minority of French Muslim women. Secularism is thus a tool of domination and formation of selves within society, and not the pure and empty form of society it, at times, has claimed to be.

In sharp distinction to this form of the secular (secularism), which is always a form of philosophical decision, a form of separating out what falls within the realm of "proper" and what is "foreign," the generic secular is an operation of entering into any form of religious thought and practice whatsoever in order to mutate it, to perform an operation of "generic forcing" that unleashes what is most true within that thought, which is to say most immanent, most in-person, or most generic. This notion of "generic forcing" is described by Laruelle as "the concrete act of the generic matrix that exceeds by their fusion or superposition (we no longer want to speak of their "unity") the duality of philosophical opposites, their division and their unity."[28] That is to say, the generic secular exceeds the duality of dominant and minority culture, of dominant and minority tradition, both as division and unity, in order to think them both as part of a larger dyad of the Real and thought (which is ultimately foreclosed to the Real). From that perspective they are superpositioned (another name for the immanence of a unified theory in *Philosophie non-standard*), they inhabit the same space as two. Neither of the two terms has dominance over the other, but each in fact is now just material that can be experimented with within the generic matrix of non-theology. Thus, what is true within any theological system will be what is most generic, what remains working after this superposition.

Non-Theology begins from the perspective of victims, but as subjects-in-struggle, it is haunted by the image of violence, always a violence that outstrips its identity as religious or secular. But if it is haunted by these images it is also productive of positive images of the generic secular. The figures in these images remain weak, minimal, in the eyes of the authorities whose use of transcendence is always in the hopes of subjugating man-in-person, of forcing the human to be unthought, to instead divert thought through transcendental circuits that disempower both theory and practice. But that weakness, that minimal character of these figures, indeed, their very generic appearance is the manifestation of the Real itself. If so often we are now required to call forth the image of terrorist violence, whether that image be the so-called Islamist violence of 9/11 or the capitalist and neoimperialist violence of Iraq and Afghanistan, then as an image of the generic secular I want to call to mind an image that follows one of those violent images. On January 1, 2011, in Alexandria, Egypt, a suicide bomber attacked those attending a Coptic church leaving 21 people dead

108   A Non-Philosophical Theory of Nature

and nearly 90 injured. This event preceded the Coptic Orthodox Church's celebration of Christmas and during the Christmas Eve service, in an act of solidarity, Egyptian Muslims surrounded the church to provide a human shield and assure their Christian compatriots of their safety and liberty to worship. The next image is more recent, perhaps more melancholy considering even more recent events in Egypt, and sprung radically into reality during the recent uprising and attempts at an ongoing revolution in Egypt. During the tense standoff with police, many of whom were corrupt and largely loyal to former President Mubarak, many of the occupiers of Tahir Square still wanted to carry out their religious duties and when the call to prayer rung out from the minarets throughout Cairo they would make their bodies vulnerable and pray. During the early days of the struggle the police would heap abuse upon the occupiers and often attack them (there is an especially inspiring video of an old woman shaming a police officer for acting without any sense of piety). To protect their Muslim compatriots during these vulnerable periods and to give them the dignity and peace required for prayer, the Coptic Christians of Egypt formed a ring around those at prayer repeating the act of becoming human shields.

What do these emotive images have to do with this project though? What do they really have to do with the generic as presented here? In short, these images are manifestations of the Real; they are manifestations of the secular messianity of the Future Christ as laid out by Laruelle. The generic helps us get a grip on how this messianity is made to function in a secular way within the theoretical practice of non-theology. Both messianity and the secular are determined in the last instance (or find their "last-identity") for and in non-philosophy by the generic. In *Future Christ* Laruelle uses the term "minimal" instead of generic, but the function is ultimately the same as minimal is a way of talking about the generic determination of Christianity as material. This is what he means when he writes poetically that "[t]he Future Christ rather signifies that each man is a Christ-organon, that is to say, of course, the Messiah, but simple and unique once each time. This is a minimal Christianity. We the Without-religion, the Without-church, the heretics of the future, we are, each-and-everyone, a Christ or Messiah."[29] While he is speaking specifically here of Christianity, theoretically (even if Laruelle himself does not bear this out in his own practice) there could be a cacophony, though simple, of minimal forms of religious material. A minimal Judaism. A minimal Islam. Even a minimal Voodoo. Each time placed within a generic matrix; not reduced to its "essence" but thought as a simple and thus radical immanence. The generic functions as a matrix within which thought develops; a generic matrix provides certain determinations for thought (the matrix itself is determined not by a meta-matrix, but by its in-One character).

NON-THEOLOGICAL SUPPLEMENT   109

Laruelle's formulation of this generic, which characterizes this matrix, is derived from philosophical materials. The importance of the generic for non-philosophy has only recently come to the forefront and comes to replace the idea of "minimal," which is more operative in *Future Christ*. Laruelle tells us in his recent works that he derives the generic from Feuerbach-Marx-Badiou. What's important in each of these philosophical constructions is the connection between humanity and science, a connection or more accurately "idempotence" that is thought more radically immanental in Laruelle. The importance of the generic in Feuerbach is largely lost to us in the Anglophone world since *Gattungswesen* is usually translated in English as "species-being." However, the French translation captures this as *être générique*. Now Marx takes up this formulation of the generic and it is Marx, Laruelle claims, that truly initiates the generic science-thought that thinks scientifically from the universality of the human. Though this remains too close to Hegel, it is Marx's presupposition or axiom of human universality that allows for his freeing of philosophy by way of a fusion with science. You have something similar happen in Badiou, except he largely returns to the pre-Marxist notion of Feuerbach's generic humanity and thinks it alongside of the mathematician Paul Cohen's concept of the generic subset (from which he takes his concept of "forcing" that is also important in Laruelle's recent work). This form of the generic undergirds Badiou's understanding of the genericity of truth procedures. Zachery Luke Fraser has summarized the five traits of this genericity as having an (1) an *indiscernible, unpredictable*, and *aleatory* character; (2) *infinitude*; (3) *excrescence* relative to the situation; (4) a *situatedness*; and (5) *universality*.[30]

Laruelle would affirm each of these, though with certain qualifications and with pain taken to secure these notions from simply falling back into their philosophical overdeterminations. For example, infinitude in Cohen refers to a *possibility* not "actual infinity" so to speak, and thus the universality operative for the generic is actually thought closer to Marx than Cohen-Badiou, for the universality of non-philosophy's genericity can't operate in the usual philosophical way where a universal is posited on the basis of its "to-come" status in the middle of a nonuniversal space. Marx himself doesn't truly avoid this with his conception of the proletariat, but he does locate in the notion of struggle something that is *futural*, but future as immanence of the human-in-person. The universality of struggle is the determination-in-the-last-identity of the human, the identity that is immanental and which other identities exist as subject to. This sense of struggle that is generic is what exceeds the specific and the individual (this is what "forcing" does in mathematics) and transforms the transcendental into simple material.

From the Generic Quantum to Generic Ecology

Laruelle's latest project marks his most complete form of non-philosophy to date. His abiding interest was always in bringing together science and philosophy into a unified theory and through his experimentations with particular religious materials, creating various unified theories of Gnosticism and Christianity, he found a model for bringing philosophy into dialogue with a specific form of science. *Philosophie non-standard* marks Laruelle's most ambitious project, the one that aims for a true philo-fiction where a human being is unafraid of putting forth a theory that respects none of the boundaries said to exist by any form of authority, scientific, religious, or philosophical. Over these three chapters I have tried to show the development of non-philosophy from its first discovery of philosophy's self-sufficiency and its first attempts to escape from the trap of that narcissistic reflection and all the way through to its present mature formulation. It is the philo-fiction of a Real foreclosed to thought, a Real before which all forms of thought are equal in this foreclosure, and which provides the general axioms that allow for an interaction with ecology that moves beyond the stale practice of either superimposing an ethical formulation on scientific practice or of wildly speculating on the underlying ontological and metaphysical meaning of nature separate from the earthly concerns of ethics.

Non-Philosophy provides both a critique of the usual philosophically imposed division of labor between itself and the sciences, and a model of thought that engages with a particular science to create new ideas. Laruelle favors quantum physics without falling into the temptation of subsuming this scientific practice into preexisting philosophical models such as materialism or spiritualism. The science is allowed to truly mutate the philosophy and the philosophy to mutate the science until they become a form of unified theory. In parts III and IV of this book I will do the same, but with a different science, with different particular material, that of ecology, combined with both philosophy and theology. This will take the form of a "generic ecology" or "immanental ecology," which makes generic six concepts derived from scientific ecology. These concepts will be the focus of part III, which will then be put to use in part IV in a theory of nature from the perspective of a unified theory of philosophical theology and ecology.

Part III

Immanental Ecology and Ecologies (of) Thought

Chapter 8

Real Ecosystems (of) Thought

Ecosystems (of) thought are real. As such ideas can be explored using the concepts operative in scientific ecology. Rather than treating the works of philosophers and theologians as if they were words from an oracle, one treats them as if their thought were an ecosystem. Among philosophical work there are populations that interact with one another (to name two dominant populations (of) thought, Being and Alterity) in a way that either creates a healthy ecosystem (of) thought, called biodiversity in ecology, or where a dominant species degrades the health of the ecosystem by spreading and destroying the niches allowed other populations. Laruelle's non-philosophy claims that philosophy always creates a united dualism, or a dualism that is ultimately united in the form of a philosophical decision, but a philosophical work demands more than this simply unilateral duality in order to operate. There are other populations (of) thought that both support this dualism of dominant species and that populate the philosophical field as the dualism itself has needs that allow for the formation of niches within the ecosystem (of) thought. Thus there is no account in Heidegger of Being without a whole host of other populations (of) thought that in turn affect that account within the unified ecosystem (of) thought. Or, to use another example, there is no thought of God in Aquinas without other populations (of) thought such as causality and Roman Catholic Church doctrine. How though do these populations interact with one another and what population can be removed from an ecosystem (of) thought while retaining its particular vitality when proposed in a different ecosystem (of) thought?

Every philosophy is built upon some never-living element that in ecology forms the inorganic spatial and temporal element of the ecosystem. Often philosophy, especially philosophy of nature, focuses on this

114   A NON-PHILOSOPHICAL THEORY OF NATURE

never-living element confusing it with transcendence or some transcendent element of things in the world. One thinks of the place accorded to logic or time in philosophical investigations where these aspects tend to be seen as dominant over and above the unified aspect of its interactions with the living and the dead. However, an immanental ecology allows us to see these elements as what they are—elements within a wider immanent system of thought. System here speaks only to the identity of this immanence, rather than its determination. Now of course these elements are necessary for the entire working of the system itself, but their overall shape also comes to be changed by the overall working of the system. There is no dominant relationship here, but only a unified working of the system as such. This is true also of philosophy and can be shown if we consider Deleuze's conception of immanence. On the surface Deleuze's pure immanence appears to be a variant of Anglo-American naturalism, or the idea that all transcendental and eternal ideas of reality should be rejected in favor of taking things as merely given and valueless. However, this is in itself a kind of transcendent idea concerning value. Deleuze's attempt to create a philosophy of immanence locates this issue, albeit not completely, and shows that immanence is never merely given but is produced or in Laruelle's terminology "lived."[1] These sorts of never-living aspects must change their spatial and temporal configurations in response to the energy exchange of the living and the dead that plays out across them and in response to collisions with other never-living elements and vice versa. Whatever is taken as transcendent and a condition of thought in a philosophical ecosystem (of) thought is always changed in the unified working of the ecosystem itself. A difficulty remaining for an immanental ecology is separating the populations that inhabit this never-living space and time and the never-living element of the populations. The way being appears in different philosophies will differ, being in one a population and in other an instance of the never-living. Teasing out this difference will vary from philosophy to philosophy and will depend upon the way the elements of each ecosystem (of) thought are discovered through an immanental ecology rather than some illusory transcendental essence of the never-living.

After the work of Deleuze and Guattari many will already accept that every philosophy also generates its own form of energy flow. What an immanental ecology does is begin to think about those energy flows more intentionally in philosophy by locating them between living thoughts and those that die on the page of philosophical treatises. Simply stated no philosopher's thoughts live on the page. Rather these dead thoughts are reserves of energy that can be consumed and thereby exchanged with living thoughts. The ecological definition of energy, following the definition given it by thermodynamics, is "the ability to do work," and a dead thought

REAL ECOSYSTEMS (OF) THOUGHT   115

is by its very disembodied existence unable to do work.[2] Work can only be done when the energy present in the dead thought is realized in some new thought. In this way a particular population (of) thought may perpetuate itself, so that when Thomists or Hegelians produce work on Aquinas or Hegel they are perpetuating certain ecosystems (of) thought, but a different population (of) thought, one even antagonistic, can also feed upon the dead thought and produce work from it that is creative within a different and possibly new ecosystem (of) thought outside of a Thomist or Hegelian ecosystem. An immanental ecology, taking these aspects of flows as given, can begin to think about the energy exchange in a constructive manner. What dead elements of past philosophies are the most productive for new and necessary ecosystems within a unified theory of philosophy and ecology?

Finally, we must begin to think about the resilience of philosophical ecosystems (of) thought. Walker and Salt give a very simple definition of resilience as "the capacity of a system to absorb disturbance and still retain its basic function and structure."[3] Why is it that by the standards of radical philosophy and within small time-scales the worst philosophies are the most resilient? The philosophies, like individualism, that undergird so many of the destructive ideologies of our age seem within the wider culture to be the most resilient to disturbance. This realization is in itself a disturbance to the radical philosopher, yet the plurality of ecosystems (of) thought from Thomism to Heideggarianism that themselves retain their basic function and shared structure as minor ecosystems within the wider ideological field must also demand our attention as well as the possibility that populations (of) thought from ecosystems like individualism may also disrupt these minor ecosystems. By understanding what makes a philosophy resilient we can begin to understand how to create a resilient unified theory of philosophy and ecology to respond to the shared problems of ecology and philosophy that are the shared problems of humanity and the nonhuman in the biosphere.

In the preceding chapters I have sketched out by way of a typology the standard relationship between philosophy/theology and ecology. I then turned to François Laruelle's non-philosophy in order to sketch out a non-standard practice of philosophy that works by thinking from radical immanence. Laruelle's method is important for this project because it envisions a different way of thinking philosophy and science, one that creates a democracy (of) thought where all forms of thinking are equivalent with regard to the Real-One, which has absolute autonomy from thought as such. In this chapter I will begin the process of constructing the unified theory of philosophical theology and ecology called for in the first chapter and that follows in the wake of Laruelle's practice of non-philosophy outlined in the second chapter.

116   A Non-Philosophical Theory of Nature

The ecology outlined here will be thought through the generic matrix of non-philosophy. In order to avoid ecologism, a kind of positivism of ecology, scientific ecology must be taken as immanental, a word I have been using in the work and that I think is immediately understandable, but I will explain it in more detail now. In his earlier work Laruelle describes the "transcendental" as "rigorously immanent."[4] In *Philosophie non-standard* Laruelle sees fit to simply coin a new term, "immanental," which describes the "non-relation" between immanence as such and experience.[5] It is then a posture of thought, like the transcendental, but one that happens within (hence the nonrelation) the experience and immanence itself. My earliest description of this project began by calling it a "transcendental ecology" analogous to Gilles Deleuze's "transcendental empiricism" or Laruelle's early description of non-philosophy as "transcendental realism" and "transcendental axiomatics."[6] The point of this name was always to refer to an idea of ecology expanded beyond its local practice and applied to philosophical theology, rather than subsuming it within philosophical theology. When the term "immanental ecology" is used, it should be kept in mind that it still shares certain likenesses with these "transcendental" positions, but the term "immanental" is more precise despite being a neologism because it already refers to the posture ecology as science takes toward the Real-One and its status as Stranger to philosophy, rather than Other or the Same. Thus it remains relatively transcendent (to philosophical theology), but rigorously immanent to the Real.

The point of the construction of a unified theory of philosophical theology and ecology by way of this immanental ecology is not an ecologism, which would give a veneer of objectivity and scientific rigor to some new philosophical conception of nature in exactly the way philosophy has always exploited science as discussed in the previous chapter, but is rather to practice a kind of "under-determination" of philosophical theology that *frees* it to think nature as a first name of the Real. It does not aim to give science a concept it requires or to force philosophical theology to recognize that it is not as respectable as science. The aim is to create a way of thinking about an abstract concept, nature, in the light of a unified theory abstracted from both the particularities of science and philosophical theology. So, with this in mind, chapter 9 outlines six fundamental conceptual elements of scientific ecology that bear on philosophical and theological thought: Populations, or the diversity of species that populate the ecosystem (biodiversity); ecological niches, which both allow for the stability of ecosystems as well as the possibility of change; the external energy relations of exchange that arise out of the populations interaction with one another and is originally provided by the sun; the never-living space and temporality of the environment; and, finally, the ecological

REAL ECOSYSTEMS (OF) THOUGHT   117

understanding of resilience of populations and ultimately the particular
ecosystem itself. Each concept is derived from the underlying fundamental
concept of ecology, the ecosystem, which I understand to be a description
of immanence.

Before turning to this immanental ecology we must first address in this
chapter a potential criticism from non-philosophy itself regarding ecology.
Laruelle is clear in his work that not every scientific practice frees conceptual thought and, though he has not addressed ecology at length, he
does offer some critical remarks regarding it that suggest ecology remains
"too philosophical" in itself. After addressing the suitability of ecology
in general and the explication of the six main principles and concepts of
immanental ecology, which form the bulk of part III, I will then consider
in chapter 10 two thinkers whose work is closest to the project outlined
here—Bruno Latour and Timothy Morton. Both Latour and Morton
mark attempts to reconcile philosophical thought with a deep engagement
with ecology (in Latour's case this is political ecology and in Morton's case
it is a more unified understanding of ecology as "the ecological thought").
In both cases it is suggested that we must rid ourselves of the concept of
nature. Not just the standard philosophical conception, but we must *really
be done* with nature for the sake of an ecological philosophy. As our project
aims to free philosophical theology by way of an immanental ecology to
recast nature non-philosophically, this is an obvious challenge.

Laruelle on Ecology

Rocco Gangle deftly captures the power of Laruelle's non-philosophy when
he writes, "François Laruelle's non-philosophy marks a bold attempt to
think the One, or Real outside of any correlation with Being and without
reference to transcendence. It is an arduous and painstaking theoretical
enterprise that must skirt the twin dangers of positivism on the hand and
false transcendentalism on the other."[7] In other words, Laruelle must navigate both scientism, or the erstwhile philosophical projection of science,
and philosophy that takes itself as the guardian of thought—philosophy
that takes itself as that which provides thought for science. As I have
already discussed at length in the preceding chapter this leads Laruelle to
practice various "unified theories" where philosophy is introduced to various other practices of thought. The goal in these dual introductions is not
to overdetermine the unphilosophical material (science, religion, etc.) with
philosophy, but to challenge philosophy through the introduction and to
treat both as simple material for thought.

118   A NON-PHILOSOPHICAL THEORY OF NATURE

The relationship with science is somewhat different though as science is both treated as material and is materially a *posture* that thought takes. The second aspect is the immanental aspect of science, insofar as it thinks from the Real rather than attempting to circumscribe and affect the Real, again discussed at length in the previous chapter. The first part, however, has special bearing here as it deals with the particular ideas and concepts operative in particular sciences and their relationship to non-philosophy. The goal of non-philosophy's thinking of the Real is always to free thought from the boundaries placed on it by specular forms of thought by, perhaps counterintuitively, locating the radical autonomy of the Real from thought (discussed in the last chapter as unilateral duality). With this in mind alongside the understanding of the generic identity of science as posture, we can see that not every science provides particular and specific forms of thought for freeing a non-standard philosophy, a wild thought (which is artificial as it is natural). Laruelle himself asks the question, "But is every science able to be utilized for this ultra-critical liberation of philosophy?" and answers, with obvious reference to Badiou, "Not every science is liberating for conceptual thought, for example set-theoretical mathematics seems to be by nature rather authoritarian, closed, and reinforces the sufficiency of philosophy, which then dreams of fiction only at its margins, a little bit like Plato."[8] The reference here to fiction is, again, a reference to the freeing of thought as practice in a philo-fiction, but what is important, again, is that Laruelle is able to recognize the need for an organon of selection with regard to scientific material.

In Laruelle's *Philosophie non-standard* the material that Laruelle thinks with philosophy as a conceptual idempotence, where two separate thoughts like waves come together to form a genuinely new wave that is not a synthesis of the two waves but is produced by them, is quantum mechanics. According to Laruelle quantum mechanics provides a true liberation for conceptual thought because, while remaining in the scientific posture that has a privileged relationship to the Real, it also "[w]eakens and disempowers philosophical sufficiency in order to free its power of invention [*pouvoir d'invention*]."[9] One of the reasons that Laruelle is critical of Badiou's use of set-theory is because it replaces the Principle of Sufficient Philosophy with a Principle of Sufficient Mathematics. Instead of freeing thought, Badiou casts a metaphilosophy where philosophy may not be able to produce truths, but it alone thinks them across the multiple terrains of knowledge. Or, while Badiou argues that we must not suture philosophy to any particular truth-procedure, he nevertheless sutures Being to mathematics as revealing the Real of Being and thereby sutures the Real yet again to Being. The generic science that non-standard philosophy aims to be requires scientific

REAL ECOSYSTEMS (OF) THOUGHT   119

material that underdetermines philosophy, again in the manner already
discussed.

So does scientific ecology meet this test or is it already too philosophical? Does it have its own Principle of Sufficient Eco-logic? It would seem
that political ecology does provide this authoritarian, closed, reinforcing of
a kind of philosophical sufficiency. Oftentimes in popular discourse this
is the role that political ecology takes in the minds of some self-styled ecological philosophers and theologians, similar to Latour's understanding of
capital "S" Science that is mistakenly taken to provide the objective end
to deliberation. The French ecologist Christian Lévêque points to the difficult relationship between political ecology and scientific ecology writing,
"[T]here is in theory no tight division between scientific ecology and activ-
ist ecology. But there is an obvious risk of confusing the philosophical
reflection with the ecological science itself."[10] When reading the various
histories of ecology, both scientific and general or political, it becomes clear
that many thinkers often fall prey to this risk and so often ecological scientists can fall prey to a kind of "science of management" that obscures its true
scientific character as a science of knowledge.[11] It is this element of political
ecology that Žižek, despite his underlying ignorance of scientific ecology,
rightly challenged in his popular polemic of 2008 where he wrote:

> This ecology of fear has all the chances of developing into the predominant
> form of ideology of global capitalism, a new opium for the masses replacing
> the declining religion: it takes over the old religion's fundamental function,
> that of putting on an unquestionable authority which can impose limits.
> The lesson this ecology is constantly hammering is our finitude: we are not
> Cartesian subjects extracted from reality, we are finite beings embedded in
> a bio-sphere which vastly transgresses our horizon. In our exploitation of
> natural resources, we are borrowing from the future, so one should treat our
> Earth with respect, as something ultimately Sacred, something that should
> not be unveiled totally, that should and will forever remain a Mystery, a
> power we should trust, not dominate. While we cannot gain full mastery
> over our bio-sphere, it is unfortunately in our power to derail it, to disturb
> its balance so that it will run amok, swiping us away in the process. This is
> why, although ecologists are all the time demanding that we change radi-
> cally our way of life, underlying this demand is its opposite, a deep distrust
> of change, of development, of progress: every radical change can have the
> unintended consequence of triggering a catastrophe.[12]

I suspect that Laruelle too is distrustful of political ecology, though the few
places he does mention ecology his true evaluation remains ambiguous. I
was able a pose a question to him that brought up ecology and to which he
did respond directly in Rome as he gave one of the keynote lectures at the

120   A NON-PHILOSOPHICAL THEORY OF NATURE

Grandeur of Reason conference in 2008. My question dealt with what I took at that time to be the underlying humanism of non-philosophy and I asked if ecology does not challenge the centrality of the human for thinking. His response, while not differentiating between scientific and political ecology, as it was a verbal back and forth, suggested that he saw ecology as performing a similar debasement of humanity paralleled in philosophy's obsession with thinking beyond the human. So, in a project that bases itself on the thought developed by Laruelle, what argument can I give that scientific ecology does provide not only the posture but also certain useful material that underdetermines philosophy and thus liberates it conceptually to think nature?

While part of the underlying background of part III will be to respond to this question in detail and show the importance of thinking with scientific ecology, I can still provide here a general argument before moving on to discuss the challenge to philosophy that comes by introducing it to scientific ecology. To do so I will first consider the only two sustained discussions of ecology I know of in Laruelle's work. The first comes to us not as one of his published pieces but as one of the occasional "Non-Philosophical Letters" that he has posted on the website for the Organization Non-Philosophique Internationale entitled "L'impossible fondation d'une écologie de l'océan." The letter, published on May 7, 2008, performs a thought experiment taking the common metaphor of philosophy as a dangerous sea and the philosopher as he who navigates that sea or the fisherman who fishes from it (found in Leibniz, Kant, and Nietzsche most famously) as its starting point. There is of course an obvious problem with this metaphor for the non-philosopher since the philosopher takes himself to be above the dangerous ocean, suggesting that there is a kind of foundation for an ecology separated from that ocean itself. In contradistinction the non-philosopher takes herself to be the boat: "Her posture (if we can put it this way) is that of a boat, and so her being-in-the-water can no longer be a being-in-the-world."[13] This will bear on his final remarks on the impossibility of the philosophical foundation in a rigorously immanent ecology of the ocean, but there is a less obvious problem and one that connects directly to his idempotence of philosophy and quantum mechanics in *Philosophie non-standard*.

Philosophy, Laruelle says, thinks in the posture of an element. It privileges thinking then from the dirt (called earth usually) or sometimes as fire, and this is reflected in its "corpuscular" posture tied to old forms of physics. Non-standard philosophy thinks according to the undulatory character of the waves and so the sea (rather than simply water) becomes an interesting metaphor-element to think from, though it should be noted that soil has a certain "wave like" quality as well. Instead of being tied

REAL ECOSYSTEMS (OF) THOUGHT   121

to a corpuscular earth, secure in our foundations, or burning ourselves up in a divine fire, the non-philosopher sets out with wild abandon on the sea. This wild abandon renounces any claim to foundation, to the idea that the philosopher owns some bit of the earth, but instead that they are in-the-water without property rights, without ideational security: "It is against 'foundation' and other similar notions as transcendent idols against which we oppose the immanence of energy or the energy of immanence."[14]

This then is where Laruelle's seeming distrust of ecology stems from. Does it as a science engage in the same kind of philosophical idol-making as those philosophers who tie themselves to a secure foundation? Laruelle ends the article by calling for a "human ecology," a remark that might seem to parallel Pope Benedict XVI's call to focus on human ecology after which the environment will benefit. This, however, is not the meaning behind Laruelle's use of the phrase "human ecology." Rather it speaks to a more rigorously immanent understanding of ecology that is called forth, but not developed, by Laruelle. To understand this better consider the final remarks of the essay. Laruelle first begins with his survey of the "situation of ecology": "Ecology's situation is as always theoretically divided between philosophies that metaphorize *physis*, theologize it as a transcendent entity of 'Nature' [*lanature*], and the physico-chemical sciences, free in themselves, which inevitably break it up. Between all of them there are the juridico-political ideologies of the 'ecologists.'"[15] While Laruelle does not demonstrate a particularly strong understanding of the specifics of scientific ecology, this does suggest that he nevertheless accurately understands how ecology functions in philosophy, theology, and as distributed among a number of other scientific disciplines.

Laruelle suggests that a more unified form of ecology could be brought about by way of non-philosophy's "last instance":

A human ecology in-the-last-instance will be theoretically more rigorous. As the man of the Last Instance is never a foundation, he must renounce or give up every "earthly" or "land-owning" foundation of an ecology of the ocean and start thinking the sea not as such but from itself, *according to the sea which is also human in the way which the human is every Last Instance.*[16]

The meaning of "human ecology" then refers to the particular immanence of man (as species-being) that non-philosophy has tried to think from its inception, rather than measuring the worth of things according to a transcendent notion of Man (what Laruelle would call the-Man): "Man can finally see his fixed and moving image, his intimate openness as the greatest secret in the ocean. 'Free men always cherish the sea…'"[17]

122   A NON-PHILOSOPHICAL THEORY OF NATURE

As I will make clearer in the next section, the purpose of engaging with scientific ecology is not simply to accept its concepts and ideas as if the project was simply a kind of ecological positivism. Rather, the task is to think infect philosophical and theological thinking on nature with certain ecological concepts that will free philosophical theology to think nature and free our thinking from its Greek overdetermination as *physis* and from its monotheistic theological determination as transcendent Other or as the simple conceptual inverse of God (this will be discussed at length in the following chapter). But not because ecology thinks nature better than philosophy, the thinking of nature still largely belongs to the province of a philosophical theology while, as Lévêque says, "ecology can no longer be a reflection of nature."[18] Rather philosophico-theological thinking will itself "go under" an immanental ecology of thought, which, in turning away from thinking moored to a transcendent notion of the-Earth, will free it to think nature in a nonreductive and nonspecular way. As Laruelle writes, "Nature is given an other-than-reductive meaning in this impossible onto-logical foundation and/or that physical powerlessness in this giving does not have definitive limitations but inhuman misunderstandings or disoriented interpretations."[19] We can change the way we understand nature philosophically and theologically by thinking from the foundation-less posture of a scientific posture.

Laruelle has affirmed much of this in a recent lecture delivered in late 2012 in both London and New York entitled "The Degrowth of Philosophy: Toward a Generic Ecology." There we see him implicitly affirming many of the ideas presented in this work, formost among them the notion that philosophy and other forms of knowledge can be treated ecologically. Of course, he states this in a slightly more polemical manner writing, "It is not a question of a 'philosophy of degrowth,' such as we sometimes hear of today, but of the degrowth of philosophy itself."[20] In this essay Laruelle goes on to try and think ecology in the "quantum spirit." This is clearly not from the perspective of scientific ecology the challenge we are putting forth for theology and philosophy, but as a kind of political ecology still, and yet this quantum notion is not far from the spirit of actual scientific ecology as I present it here. Laruelle writes,

> The quantum model obliges me to maintain the correlate of physics (physics as essential quantum, not the traditional *physis*) is the universe, and not the world. I understand by "universe" the correlate of modern knowledge, by "world" the correlate of philosophy. I would add that the universe is not the great mystical All evoked by certain physicists, but an epistemological correlate of physico-mathematical knowledge. The universe, even as an object of experimentaiton and above all if it is an object of experimentaiton, is an object of knowledge, not a material object.[21]

REAL ECOSYSTEMS (OF) THOUGHT   123

This notion of a universe that is not the great mystical All maps neatly onto the notion of nature presented here. For nature will turn out not to be a physical object, not an All (this is the hallucinatory theological form of Nature explored later), but remains where the relations of the ecologies (of) thought play out. Laruelle also seems to be thinking something similar when he writes, "Let's suppose an ecology of the relations of thought, of its highest forms of which we can make use—science and philosophy, art and religion, relations with and within the universe."[22]

Yet the lingering problem of the temptation to humanism remains in this essay as well. Laruelle notes that "[t]o preserve the natural environment of existence, to preserve man and his survival qua species even, is the immediate and primary aim of ordinary ecology."[23] But this goal, for Laruelle, must submit to a more primary defense of human beings, "The 'defence' and the maintenance of human environments, spontaneous and naturalist ecology, must be reordered in view of a defence of generic man in (and sometimes against) the environment or milieu of knowledges."[24] Laruelle is not committing himself to a human chauvinism here, however, as he does go on to write, "This new objective of ecology cannot be called superior or meta-ecological. It is in-the-last-instance a *generic usage* or epistemic milieus, the best appropriation of knowledges (including philosophy itself) in view of the defence of humans against their self-destructive drive, which has its origin in the world."[25] In other words, the problem of human environments and even going against the environment is the same problem that haunts this work—that nature has become a problem for nature. Though Laruelle in some sense clearly is walking a thin line, one that he may not successfully traverse in future works on the relationship of the animal and the human, it is a line drawn by political ecology itself. For all of our green consciousness

> is still to presuppose that man can decide freely, in some all-powerful manner, to safeguard nature or to destroy it. Whereas he does not really have this power to transform it wholesale, since he himself belongs to every decision, is included in it and perturbs it, puts it back into play with every decision or repetition. He has only the power to underdetermine his decisions.[26]

So let us go under ecology in order to confront ourselves along with nature, to make the decision if we want to go on living or not, to live as natural creatures against the hallucinatory Nature we have projected.

Chapter 9

Elements of an Immanental Ecology

In this chapter I will describe six fundamental conceptual elements of scientific ecology after a discussion of the ecosystem that they all relate to. The point here is to think philosophically and theologically through the material of scientific ecology. But the presentation of these ideas will not simply be historical or scientific or philosophical, though they will also be these things, but presented in the same style as Laruelle's "generic science" or unified theory. This means that though the integrity of the scientific nature of the ideas is respected, meaning we don't treat them as "absolute" concepts, they are read with regard to philosophical theology rather than terrestrial or aquatic ecosystem dynamics. Whereas Laruelle brought together quantum mechanics and philosophy by way of "idemmanence" or the immanence expressed in quantum physics's conception of idempotence, we bring together ecology and philosophical theology by way of an "ecology (of) thought."[1] In this generic science thought itself is treated as if it were operative within an ecosystem. The point of this thought experiment is not to create an ecosophia, as Naess aimed to do, but to show that the normal philosophical and theological ways of thinking about nature do so without engaging with scientific material. As I showed in part I, even when scientific material is engaged with, it is never a deep engagement with *ecological* scientific material. Again, rather than a simple naturalism that ecology or other forms of science would undergird, here we begin with the idea that nature is manifest in ecosystems, such that ecology does think it in some relative way, but nature as such is not the object of ecology. Nature, as perverse, is only the object of a unified theory of philosophical theology and ecology. This doesn't exclude other unified theories, though ecology is especially useful in this instance because it is a synthetic science bringing together the sciences of biology, physics, biochemistry, geology,

126 A NON-PHILOSOPHICAL THEORY OF NATURE

climatology, and many others in its study of the ecosystem. In its own openness to disciplinary perversion it is able to express in its principles and concepts nature's perversion.

An obvious retort to this thought experiment is that this is simply a metaphor and reveals nothing interesting in itself. Of course the Greek *metaphora* literally means "a transfer" and comes from the verb *metapherein* meaning "to transfer, carrying over." The charge is that this "just a metaphor" suggests we are unduly "transferring" an idea from scientific ecology to philosophy. In short that we are not respecting the proper borders that allow the standard division of thought to work. This is also an economic division of thought, one relating in part to labor (so who does what jobs) and in part to branding. Of course there is a reasonable warning in here as well, suggesting that we should not disrespect the specificity of practices within thought. This need not be a "Sokalism," but can point to the real difficulties present in both scientific thinking and philosophical/theological thinking. Consider the outrage among theologians and scholars of religion when those trained in biology or cosmology publish their largely unlearned thoughts on religion; surely then the converse would be true as well.

So, just as Spinoza wore a signet ring with a red rose and the Latin *caute* [caution] that he used to mark his texts with, we note that we are proceeding cautiously into the scientific material. But we are still proceeding because even if it were just a matter of a simple transfer of material from scientific ecology to philosophy it could still then be following the principles of ecology at the level of fiction. Thus one cannot see it as a metaphor if by that one means something "outside but like an ecosystem," for the metaphor too depends on the ecosystem for its existence since thought is tied to actuality and actuality is ecological. One may name this the materialist element of immanental ecology, but only if what the material is remains open to revision. However, I want to suggest that while this is certainly a work of "philo-fiction," there is also a more substantial basis for positing an ecology of thought. Consider again Morton's remark that thinking is one of the things damaged by modern society in addition to actual ecosystems.[2] The environment inhabited by human beings, which to say whatever particular ecosystem different human societies have codeveloped in, is part of development of ideas. This idea was already present in the first attempt at universal history by Ibn Khaldûn, and contemporary environmental historians have shown that ideas are part of ecosystems.[3] But this is obvious already in everyday popular environmental discourse on the destructive nature of certain human ideas. If these ideas are allowed to continue they will run up against ecological limits like any other species. So, for instance, the idea of wilderness is currently on the decline because it

ELEMENTS OF AN IMMANENTAL ECOLOGY   127

leads to practices that weakened ecosystems in the American West.[4] Note
that this claim doesn't need the ecological status of ideas to be dominant
in order for it to be true. So, I am not suggesting that ideas are what are
called "keynote species" that greatly determine the ecosystem. Ideas, like
most species, are "under-determined" by their ecosystem. In other words,
we can't simply reduce ideas to their localization within an ecosystem, just
like we can't reduce a particular plant species to its same localization, but
ecosystems nevertheless force certain ideas by way of their same immanence. The ecosystem is both not the sum of its parts (so it is not reducible
to the collection of species and the exchange of energy between them) and
it is not really more than the sum of its parts either. This is why particularly ecosystems are notoriously difficult to delineate in the way that exact
and abstract mathematical sciences are able to delineate their objects of
knowledge. But for more on that let us turn our attention more intentionally toward the ecosystem.

Ecosystem

The ecosystem concept is key to scientific ecology. Within the scientific
field there tends to be a loose division between those researchers who practice ecosystem ecology and those who practice population ecology. The
divide between these two postures can be explained in part by a sociologi-
cal cause, namely, ecologists have a strong desire to be accepted by other
biologists (for ecology was and remains a biological science) as well as other
scientists in general, who often hold up physics as the paradigm of scientific work. However, due to the difficulties of studying actual, complex,
and changing systems, ecology is often not able to rise to certain epistemological norms common to more abstract sciences; for instance, falsifiability
is notoriously difficult in ecology.[5] But population ecology, or the major
subfield of ecology that concerns itself with the way populations of species are distributed in time and space and their interactions among other
populations, allows the individual ecologists to do science in a way that
is more generally accepted by the social community of scientists because
it is easier to observe and conduct experiments on discrete populations.[6]
The disciplinary reality, according to Lévêque, is that ecosystem ecology
and population ecology are developed in relatively autonomous manners.
Population ecology fits a generally reductionist program, while ecosystem
ecology attempts to understand "the cycles of matter and energy that structure the ecosystems" and can be said to be holistic in some sense.[7] There
is, though, no real reason why a population ecologist and an ecosystem

128   A Non-Philosophical Theory of Nature

ecologist wouldn't work together because populations are determined by their place in any ecosystem and ecosystems are formed in large part by the populations that populate them. So, despite this seeming split, ecology is at least virtually a unified scientific posture and this is especially true of ecosystem ecology as F. Stuart Chapin III, Pamela A. Matson, and Harold A. Mooney make clear writing, "Ecosystem ecology therefore depends on information and principles developed in physiological, evolutionary, population, and community ecology."[8]

The ecosystem concept finds its "exact moment of birth [...] when the English ecologist Arthur Tansley created the word and presented it in a technical paper."[9] As I have already shown in the first chapter, this development allowed ecology to avoid being too determined by metaphysical systems of philosophy, namely, determinism and organicism. Yet, Tansley's idea, while being taken up and used in an ad hoc manner, was not taken by most ecologists to be the grounding principle of ecology. This didn't happen until Eugene Odum's *Fundamentals of Ecology*, written in 1953 along with his brother Howard Thomas Odum, when he placed the ecosystem concept at the heart of ecological research because it is the largest functional unit in ecology.[10] Ultimately the ecological principles laid out in this work follow the Odum interpretation of ecology and so it is necessary to understand the purpose behind Odum's placement of the ecosystem concept in his book. As Golley shows in his history of the ecosystem concept, the impetus behind Odum's textbook came from a disagreement between Odum and other zoologists at the University of Georgia. In late 1940s Odum pushed for the inclusion of ecology in the curriculum but the rest of his department voted down the proposal on the basis that ecology had no specific principles in itself. Odum responded to this with a book clearly outlining the fundamental principles of ecology. The importance of the ecosystem concept, laid out within the first nine pages of the book, is summarized by the distinct ideas it leads to in ecosystem science. Golley clearly summarizes this:

> In his explanation of the ecosystem concept, Odum developed several distinct ideas. First, the largest ecosystem is the entire earth and the biosphere is that portion of the earth where the ecosystems operate. Second, ecosystems may be of various sizes, from the biosphere to the pond. Third, animals and nongreen plants are dependent upon plants that manufacture protein, carbohydrates, and fats through photosynthesis; plants are controlled by animals, and both are influenced by bacteria. Forth, organisms also influence the abiotic environment. Fifth, humans have the ability to drastically alter ecosystems. Thus, an understanding of the ecosystem concept and the realization that mankind is part of these complex biogeochemical cycles is fundamental to ecology and to human affairs generally. By page twelve of

ELEMENTS OF AN IMMANENTAL ECOLOGY   129

his textbook Odum had laid out an agenda that anticipated the direction taken by the ecological movement over the next twenty years.[11]

Golley credits Odum's textbook with turning ecology into a recognized science in its own right, which was recognized by the wider public as well as the scholarly community. In part this was because in Odum's interpretation of ecology its "language and concepts reflected some of the most advanced trends of the 1950s."[12] These advanced trends coalesced through the twentieth century into general systems theory, from which ecology continues to derive many tools.

It was in part Odum's recasting of the ecosystem concept along the lines of general system theory that recovered its importance to a new generation of ecologists. Odum recognized the ecosystem as both a biological unit (like an organ or organism), suggesting a determined static corpuscular idea, and also that which is made up of the biotic community and its environment, suggesting something more akin to a wave idea.[13] For the scientists this fundamental ambiguity or duality, found also in Tansley, has been a source of distress as it continues difficulties ecologists have found in trying to make their science more deterministic, in part to live up to the ideals of science set by classical physics. Even though, again, classical physics doesn't have to deal with the same level of dynamic, living complexity of systems that ecology does. Here I hazard a thesis that will no doubt be controversial to scientists; the fundamental duality of an ecosystem is also its strength in thought. The duality of the ecosystem between discrete unit and dynamic flux finds an analogue in the wave/particle duality of matter found in quantum physics. In short, looked at from the perspective of a biological unit, the ecosystem is a unit, and from the perspective of succession, the ecosystem is made up of the biotic community and its environment.

In the ecologists' attempt to describe the Real, they have been led to think the ecosystem concept. The concept was originally formulated to avoid philosophical and theological overdetermination (as discussed in the first chapter), but this ultimately dual character of the ecosystem as both discrete unit and flow point to an expression of immanence within this conception of the ecosystem. The ecosystem concept is a post-Newtonian concept that is determined more by an undulatory or wave-based model than a corpuscular or particle-based one, in opposition to Laruelle's mischaracterization of ecology by collapsing it into standard philosophical visions of ecology. The ocean is not the site of an impossible ecology, but is quite simply the research-world as ecosystem where ecology thinks. Ecology is concerned with the flow of energy between the living, the dead, and the never-living. These are not distinct others, but are the ecosystem's

130   A Non-Philosophical Theory of Nature

identity in its radical immanence. Under the posture of ecology a dead worm and a living bird = One while remaining what they are.

Adding support to this casting of ecology as expressing an undulatory character is the special place of aquatic biology in the development of the ecosystem concept. Prior to Tansley's formulation of the ecosystem concept Stephen Alfred Forbes posited a similar idea in 1887 that was ignored, Golley suggests, in part because it was published in a minor regional journal with readers mostly in central Illinois, United States. Forbes studied lakes as "microcosms" of a wider "community of interest, between predator and prey."[14] Golley suggests that Forbes's idea, taken up by limnologists, advances beyond Tansley's later and more popular conception because it aims for the development of a functional approach requiring extensive knowledge of what organisms live within the system and what links them before scientist can understand their organization as a system.[15] Perhaps owing to the later influence of Forbes by way of later limnologists after the popularization of Tansley's idea, lakes become the site of the first advanced ecological studies that were able to examine and express formally the flows of energy between species and their living and nonliving environment. Through these studies ecologists could observe the time of a particular ecosystem, this is called succession, and consider succession as part of the function of the biota.[16]

The obvious reason that lakes were so important for the development of ecology is that a lake has relatively obvious boundaries with other surrounding ecosystems. Yet, this should still not suggest a corpuscular-based theory of ecosystems or even an organicist one, but rather simply a practical need to find a physical body to study from which the principles of ecology could then be developed. Ecology is decidedly an actualist science, in that it must work with actual systems and derive all its experimental theories from that system, and so it is perhaps more fundamentally immanentist than a largely virtual science like quantum physics, which can engage in more abstract and free-floating thought experiments.

I have already touched on the ways in which ecology as a science has avoided being overdetermined by philosophy and theology. As I outlined in the first chapter this attempt was more or less gravitated around the concept of the ecosystem, but it would be a grave mistake to think that this concept was given once and for all. Science does not work that way. Instead, the concept was used axiomatically; when it worked it worked and when it did not work the concept would be adjusted. This axiom-in-action is what differentiates the science of ecology from the attempted theoreticism of philosophy and theology.

This active axiomatic structure is related to the way ecology is distinct from the other scientific practices that it draws upon and that ecology

ELEMENTS OF AN IMMANENTAL ECOLOGY   131

modifies for its purposes. People often ask how ecology is different from general systems theory, how its theory of energy differs from thermodynamics, and if it is really distinct from biology in general. What is behind these questions if not a certain kind of philosophy of science that aims to find *the* science that will reveal all the required answers for a philosophy of reality. A division of the sciences that creates a permanent war between the real sciences and those sciences that depend upon them. But there is no agreement from the philosophers on which form of science is *the* science and so this war is really an internal philosophical war. What is the non-philosophical response? A relatively peaceful one, for the way non-philosophy conceives of the division of ideational labor locates all scientific practice and all philosophical thought relative before the Real. Ecology is thus only different from those scientific practices and theories it draws upon by virtue of its object, which is the ecosystem. The object toward which a science is directed is immanent to the identity and posture of that science such that the autonomy of a science is always given not by some particular practice but by the object that produces that project. In other words, the differences in the practices between theoretical set theory and ecology are entirely important with regards to what can be done with the science on different occasions, but those practices cannot be truly separated from the occasion itself. The object brings with it the problems and practices of the science. The problem of scientism arises when one tries to make a single science and its object (what Laruelle calls its "research-world") *the* form of thought above all others.[17] Ecology already resists this kind of overcoding of everything by the very virtue of the complexity of its object. For quite simply the ecosystem is not an abstraction that can be experimented in a laboratory, but is a lived, immanent reality in time.

Biodiversity

Biodiversity is a relatively recent principle in scientific ecology and it isn't without its controversies. Lévêque, for instance, prefers the term "biological diversity," even while recognizing that biodiversity is simply a contraction of this term, because of its all-too-easy appropriation by political ecology. He points out that the term was popularized in the context of the alarm of naturalists at the rabid decline and destruction of natural environments.[18] And, as we will see in more detail later, even Edward O. Wilson, who is one of these popularizers, raises issues regarding the fundamental unit of biodiversity. Regardless of these issues the scientific investigation into biodiversity is a response to the incredible diversity of life on planet Earth.

132   A NON-PHILOSOPHICAL THEORY OF NATURE

What scientists like Wilson have argued, supported by more recent work in resilience ecology, is that this diversity is necessary for and constitutive of the stable and resilient functioning of life on Earth. Moreover, it raises questions of a philosophical and theological character about the concept of nature itself. Namely, what is nature in relation to life? Is one term primary over the other? Or are they reversible with one another? Beyond these questions the field of study around biodiversity gives us tools that can be used to disempower philosophical and theological overdetermination of ecology and begin to treat thought as if it were ecological (which, of course, it always already is).

If it is true that philosophy begins in wonder (one thinks of Plato) and theology in the fear of God seen through God's Creation (one thinks of Paul), then it is worth noting the reported numbers of biodiversity as summarized in Wilson's *The Diversity of Life* (2001), which continues to be a standard summary of debates and issues in the study of biodiversity. According to estimates made by Wilson in 1986 and published in 1992 the number of living species was put at 1.4 million. He goes on to state that 13,000 "new" species are discovered each year, meaning at the time *The Diversity of Life* went into its second edition in 2001 the number had gone up to 1.5 million. Of these we have catalogued 865,000 different species of insects and 69,000 species of fungi. These are the species we have identified, the estimate of the true number of species on Earth ranges from 3,635,000 and 111,655,000. The vast majority of these unknown species are ones that don't normally occur to the human being to care about. So, while it is estimated that we know 98 percent of the living bird species, we know only 1.5 percent of chromophyte algal species. And as fantastic as these numbers are, literally suggesting a planet that is teeming with living creatures and where that very teeming provides the conditions for the remarkable creation of more species, it is estimated that 98 percent of all species that have existed on this planet in the more than 3.5 billion years of biological history have disappeared into extinction forever.[19]

Note that these numbers refer only to species. This is the unit preferred by those who study biodiversity because of the relative stability of species as a concept (defined by Wilson as "a population whose members are able to interbreed freely under natural conditions") and the relative ease in recording them.[20] For, the fact is, there are other organic units the biologist and ecologist could use in their study of biodiversity, such as genes or, at the other end of the quantitative spectrum, ecosystems. Wilson tells us,

> Biologists still find it useful to divide living diversity into a hierarchy of three levels: ecosystem, species, and gene. [...] Because ecosystems are so often hard to delimit, and because genes are difficult to identify and count,

ELEMENTS OF AN IMMANENTAL ECOLOGY 133

the unit of choice in biodiversity studies remains the species, which is relatively easy to diagnose and has moreover been the central object of research for over two hundred years.[21]

Yet, even setting aside the difficulties of counting and identifying the unit of species, there remain some enduring problems with the concept of species.

The most obvious problem relating to the conceptual description of species is simply that not all organisms are sexual and so the concept of species doesn't fully account for them. However, since the overwhelming majority of organisms are sexual in relative terms and furthermore the majority of nonsexual species have evolved from sexually reproducing ancestors, this isn't a conceptual problem that has caused scientists to give up on a concept that otherwise works.[22] What is more problematic, and interesting for a non-philosophical theory of nature, is the inconsistent nature of species. As Wilson pithily states, "For species are always evolving, which means that each one perpetually changes in relation to other species."[23] This causes a nuisance in terms of cataloguing some "sibling species," which are largely similar but nonetheless distinct, but more troubling conceptual is hybrid species that are created within a genus.[24] This is perceived by scientists like Wilson as a problem for the concept of species and moreover as the fundamental unit because it partly opens the gene pool leading to the question of the identity of species as such.

However, the problem present for the ecologist is an occasion for speculation for the non-philosopher. In both the standard definition of a species as a community where the population can freely breed and in the problem where gene pools appear to open up there is still a generic, radical identity. In the technical language of non-philosophy this radical identity is in-One, meaning it is Being and its Alterity or difference from other species is secondary to its own actuality or identity, and so what we have here is a new notion of identity than the one that Wilson is presupposing. In the non-philosophical conception of identity the issue is fundamentally related to actuality.[25] The hybrid species is radically actual, regardless of the inadequacy of the thought attempting to think it. This is an example of the fundamentally unilateral relationship between the One (which is occasional and thus not a single substance, being prior to Being, but always manifest-without-manifestation in multiple sites) where the actuality or identity of the hybrid species requires that thought think it, but thought itself does not have any absolute effect (even if it may have a relative effect) on the actuality of the hybrid. Perhaps, then, a non-philosophical solution to these enduring ecological problems could be offered by way of the unilateral duality of identity. Keeping in mind that there is always the absolute unilateral duality of the Real-One and some clone or identity circumscribed in thought, in the instance of the hybrid there is a duality between the hybrid-species and the closed-species. Which is primary, meaning which provides the radical identity for the concept of species? It may seem, considering the requirement given by ecology for some stable fundamental unit, that it is the closed-species that is primary. However, this would be a quick retreat into the realms of epistemology (which scientific practice is autonomous from) because it requires that ecology pass a philosophical test regarding a closed conception of identity where stability comes to overdetermine identity as such.

It is not the closed-species that is primary but the hybrid-species because it is the hybrid-species that is the productive force at work in biodiversity. Consider Wilson's remarks against the notion of a chaotic notion of species inherent in discarding the closed-species as fundamental:

> [R]eproductive isolation between breeding populations is the point of no return in the creation of biological diversity. During the earliest stages of divergence, there may be less difference between the two species than exists as variation within them. A surge of hybrids may yet occur to erase the barrier and confuse the picture even more. But in most cases the two species are embarked on an endless journey that will carry them further and further apart. The differences between them will carry them further and further apart. The differences between them will in time far exceed anything possible among the members of their own breeding populations. In the real world, the great range of biological diversity has been generated by the divergence of species that were created in turn by the defining step expressed in the biological-species [what I have been called closed-species].[26]

While Wilson feels it necessary to protect the closed-species concept, in part because of other population biologists who are "enchanted by the dynamism of the speciation process and the many problems thrown up to embarrass the biological-species concept in the early stages of species separation," this doesn't appear necessary from a conceptual standpoint. For, as Wilson himself states, reproductive isolation is the point of no return in the creation of biological diversity, but this isolation is yet subject to the time it takes for the species, including hybrid ones, to diverge from each other.

Identity in the case of species could be said to be in-biodiversity. What is primary in the species concept is not a closed gene pool, but the drive toward diversity playing even on these closed gene pools. This gives a certain unified character to the conception of biodiversity and one that will be extended to philosophy and theology as ecosystems (of) thought. What is this unified character? It is applying the principles of biodiversity to the concept of biodiversity itself. The biosphere is, in a radically

ELEMENTS OF AN IMMANENTAL ECOLOGY   135

immanental way, both constitutive of the diversity of life on earth and constituted in both its functioning and resilience by that diversity of life. Thus species themselves are subject to the drive toward diversification as individual species. Keeping in mind that the famous conception of species-being, developed first by Feuerbach and subsequently expanded by Marx as the German *Gattungswesen*, is more closely related to the generic as already discussed. Though this relation is obscured for us now, species and generic are closely related terms. This closeness opens up the possibility of treating not just human beings as "species-beings," but all of creation itself when "species-beings" is thought through the concept of biodiversity.

This will become clearer in part IV, but I will summarize the thrust of the argument here. I have suggested that, on the basis of the acutalist and pragmatic scientific practice of population biology, we can locate the radical identity of species in the drive to diversity present in the way hybrid species crack open closed gene pools. This doesn't negate species for population biology, since as a fundamental unit of measure it works, but it does require that the fundamental unit be taken as occasional rather than absolute. In practice this is already accepted. This conception of species, derived from locating the relationship of unilateral duality operative within it, is subjected to the very principles of biodiversity. If the resilience of the biosphere as well as individual ecosystems is dependent upon biodiversity, then biodiversity is also subject to the force of diversity as well. This has consequences for understanding the way ideas function within ecosystems (of) thought, where the specular tendency of philosophers and theologians is to construct an absolute or closed system. But this is always resisted by the movement of those thoughts into a wider intellectual field where other thinkers take up these thoughts and mutate them. It then also has consequences for our philosophical and theological conceptions of nature, suggesting that the notion of nature as "birth" (the meaning of the Latin *natur*) may be salvageable, but not as some kind of hidden metastable field that deplores mutations. Rather hybrids and all sorts of perversions are birthed forth *naturally* in nature.

This should change our understanding of the natural. It utterly destroys the sense of the natural carrying with it any sense of normativity other than the normativity of immanence as such. This is an important distinction, for normativity is not completely thrown outside of thought or nature, but it is reconfigured from the position of an immanence or, what amounts to the same thing here, a unified theory of philosophical theology and ecology. This will be developed at length in the next chapter when dealing with Aquinas and Spinoza in relation to anthropomorphism, but here we can sketch out a formalist theory of the normative content of this non-naturalism. By non-naturalism, again calling on the logic of non-philosophy, I mean a conception of thinking from nature that is no longer subservient to the natural, but to a conception of nature as if present in the radical immanence of ecological concepts that are lived, such as biodiversity and the niche. Here normativity is not to be found in some nature separate from human beings, but human beings are able to derive norms from their knowledge of ecosystems. These are extremely limited insofar as they don't provide models from on high, but as norms are still relative to the absolute autonomy of the Real and thus material that can be worked with alongside other creative labors. So a formalism of the normative content of biodiversity can be expressed in this way: if, as Deleuze argues in his reading of Bergson, dualism = monism then this may also be expressed as immanence = pluralism. Deleuze's point regarding Bergson is that differences in kind (in distinction to differences in degree) are always returned to some common virtual point where these differences converge.[27]

The virtual though is not lived as such, but is produced by lived experience. Thus the virtual monism whereby the relative dualisms at work in Bergson's vision of reality converge are unilaterally determined by the pluralism of generic actualities. We may thus write a corollary formalism to the first as: pluralism = (dualism = monism) as a unilateral duality. This would mean that the pluralism of the identity of actualities, regardless of their place in duration, is foreclosed to the mixture of dualism and monism because the plurality of effects of the Real (which is One, but not as substance or Being, but only as simple identity) produces the mixture itself. What does this formalism then represent in terms of the normative content produced by biodiversity? Simply that any appeal to "the natural" will fail to find any kind of basis outside of a general affirmation of the plural productive forces of the biosphere. In order to provide an ethics the natural must itself be subsumed within a wider ecological framework where that pluralism is a kind of affect undergone by ethical thinking, but that is underdetermined by it such that it can still produce ethical statements outside of a purely naturalistic conception. In this way the natural is no longer separate from the realm of human thinking, what we might call somewhat problematically "culture," but is rightly seen as another perverse production or effect of nature via human fabulation. Or, in other words, the natural is not something absolutely transcendent to the human, but is a population (of) thought that was produced naturally just as humans were and subsists in various ecologies (of) thought, but as such is subject to the wider principles of a generic ecology and immanent to those ecosystems as such.

ELEMENTS OF AN IMMANENTAL ECOLOGY   137

Niche

Biodiversity opens up to a corollary concept that ecologists refer to as the niche. If biodiversity is the recognition that there is a principal drive to diversification within the biosphere, niche theory is the attempt to give shape to the functioning of biodiversity. For biodiversity is a principle derived from the research into the proliferation (one might even say clamor) of species that are identified by the ecologist as those populations that can freely breed under "natural" conditions. Niche theory is able to locate the ways that clamor comes into a stochastic harmony. This stochastic harmony is described by Paul S. Giller as population interaction with other populations (this grouping of populations is called the community) and the wider ecosystem.[28] Giller clarifies the strict definition of a community writing that a community is "a combination of plant, animal, and bacterial populations, interacting with one another within an environment, thus forming a distinctive living system with its own composition, structure, environmental relations, development and function."[29]

Niches are tied more closely to the community rather than the ecosystem as a whole, though again the confusion with regard to scale of ecosystems makes this a somewhat unclear point. Giller helps clarify the place of the niche when he writes, "The ecological niche is a reflection of the organism's or species' place in the community, incorporating not only tolerances to physical factors, but also interactions with other organisms."[30] In a nontechnical sense, though nonetheless true, niche refers to what lines of sustenance are open to the organism or species. That is, a niche is that place, within a network or mesh of interactions (these are always approximate analogies for the mathematical model of the energy exchange), where an organism can find enough energy to continue to live while passing on its genetic information. Now the niche of one species may be wide enough to allow that species to spread across the ecosystem, and even, as in the case of human beings, to dominate the ecosystems they exist within. This idea of domination refers to the intensity of the effects that this species has on the particular ecosystem. So the human being has obviously had a high magnitude of effects on the ecosystems they inhabit and has even shaped them. This limits the niches of other animals, while opening up other niches. If the human species were to disappear the ecosystems they had inhabited would no doubt change fundamentally, which is not necessarily true of species who have smaller niche widths.[31]

In practice most organisms and species are limited or "checked" by other organisms and species. This should not suggest a rather medieval

138   A NON-PHILOSOPHICAL THEORY OF NATURE

notion of hierarchy based on an anthropocentric understanding of power, but in ecological theory hierarchy is always more complex and open to reconceptions of power more akin to the focus on potentiality that has been somewhat common in European political philosophy since the 1970s. For bacteria, that black hole of biodiversity, may end up being a dominant species or at least one that checks the niche width of other organisms and species in a significant way. This may seem like a strange statement but it is because "in the real world" the environmental gradient (or space) where niches exist "is not measured in ordinary Euclidean dimensions but in fractal dimensions. Size depends on the span of the measuring stick or, more precisely, on the size of the foraging ambit of the organisms dwelling on the tree. In the fractal world, an entire ecosystem can exist in the plumage of a bird."[32]

Yet, even with this *n*-dimensional space of the bird's plumage or the single stick in the forest, there is always some check on the hypervolume. This check is referred to as the principle of competitive exclusion, which holds that if two or more species coexist there should be some ecological difference between them.[33] This is not an ironclad law as Wilson reminds his readers. For even though one dynasty of species cannot tolerate another dynasty of a closely similar kind and "when one group radiates into a part of the world, another group must retreat," this is only a statistical tendency that clues the ecologist in to the likelihood of some ecological diversity at work where two seemingly similar species do coexist.[34] There is something interesting at work here, which tells us something about the weakness of a crude quantitative measure with regard to dealing with the ethical issues raised by ecology, for it may seem that a species should simply be considered endangered if it has a relatively small quantitative population. Yet, it is its niche width that is really the matter of concern, such that a population can be large and even widespread, but if its niche is scarce the species resilience is weak and it is threatened. A change in the wider community structure could lead to disaster for the species.[35]

Already discussed earlier, there has been some critical noise from philosophers like Slavoj Žižek concerning the supposedly ecological ideology of "balance." What Žižek is really referring to is not ecology as such, but certain notions that are found in popular quasi-green discourse. These forms of discourse can be called quasi green because they are actually old notions, more mechanistic and theological as Botkin and others trace the history of ideas, which are easy for mediatic public intellectuals to use as well as members of the paid commentariat whose job it is to have opinions about all public matters regardless of their own lack of precision or knowledge. The concept of niche is a good example where the philosopher goes wrong with his vision, where the attention he gives is determined by

ELEMENTS OF AN IMMANENTAL ECOLOGY   139

his philosophical faith, allowing him to cast derision on the unthinking scientist, and so he may see the niche as the old philosophical idea of balance. Or take the theologian, with his own faithful attention, who may see in the niche nothing but an ontology of violence. In truth, neither balance nor ontological violence is required by the concept of the niche when it is placed in an immanental posture and extended to thought itself.

The concept of the niche has to be thought through the concept of the never-living (discussed more in the next section), rather than in the dialectic of life and death that both the philosopher and theologian persist in thinking through. While we deal more in the final chapter with the ethical problems manifest in ecological theory, for now the point need only be made that the niche concept is only part of a wider theory. It does not overdetermine the value of species without there being some hidden philosophy or theology behind it. What the niche concept does point to is a generic posture of all living organisms. Not that of violence, if by violence one means Greek *agon* or of the violence committed against the hostage, but of immanental struggle in the World as separate from the notion of a "whole." Each community is a stranger to the biosphere insofar as it can be identified as a community and if it plays its part in the functioning of the whole it does so without some kind of intentionality. The biosphere simply is the various community-identities functioning within the same *n*-dimensional space.

The niche is both a model of immanence in the ecosystem and a stumbling block to standard philosophical conceptions of Nature as read in tooth and claw, as primary qualities, and also of the New Age-y conceptions of balance. Of course, anyone would be delusional to claim that nature isn't violent, that in nature and all its perverse manifestations there isn't some affective field of sorrow and mourning. But there is also a fundamental joy and this is expressed also in the productive powers seen in the niche. Ecology breaks with philosophical naturalism insofar as the scientific practices of ecology, which are able to locate niches and biodiversity, require no simple reduction to "the natural." This is not to say that individual ecologists may not believe in some form of naturalism, but that this belief will always ultimately be external to the immediacy of their practice and even from the underlying principles and concepts of ecology. But the niche, outside of any reduction and remaining radically immanent to the lived aspect of the species, witnesses to the underlying perversity of nature. One might even say a "joyous perversity," if by that one means that the creation of the niche witnesses to the species living without regard for death.

New species come into existence. This is the protest of the immeasurable perversity of nature against the notion of Nature at work in ideas of "the natural." This form of naturalism lies at the heart of philosophy's ruminations concerning nature going back to Heraclitus's aphorism usually translated as "Nature loves to hide" but that Pierre Hadot suggests is better translated in one of two ways: "What causes things to appear tends to make them disappear (i.e., what causes birth tends to cause death). Form (or appearance) tends to disappear (i.e., what is born wants to die)."[36] Such being-toward-death has dominated philosophical conceptions of nature. But this is true of Christian theological conceptions insofar as Paul's theology of Creation, which states that "the present form of this world is passing away" (1 Cor. 7.31 NRSV) is at the heart of Christian ideas about nature. In both cases the present form of nature naturally tends toward death and this natural tendency at work over and against the living is what determines all thinking concerning nature. Nature becomes another name for the realm of violence. But nature reconceived through a non-philosophical unified theory of philosophical theology and ecology (remembering that nature is not the object of ecology, but is an effect produced from the Real, which can be thought using the tools of ecology) can move past the natural and thus past the overdetermination of the dialectic of life and death.

The niche is the production of the living against the requirement of death at work also in nature. Yet, this protest would be in vain if it simply hoped to overcome death by destroying death. Biologists have a name for the living form of this desire—they call it cancer. For cancer is simply a living cell refusing to expire, refusing the programmed death of apoptosis and thus destroying the wider system it is within. The niche is an expression of protest against the necessity of death insofar as it pays no attention to death *as such*. Death never determines the niche in the way it determines philosophical ethics or religious fantasies of overcoming death.

We can illustrate this argument by way of a creative recasting of the persona of Job, a persona that has been used both by philosophers and theologians. For if we think of the niche as a resistance to death, as a resistance to the terms set by Nature that philosophers hallucinate, we see an underlying identity common to the immanence of the niche and that of Job. What the niche shows is that we can discuss nature as perverse against the terms set by Nature, just as Job perversely stood up against the terms set by God in refusing to accept what his friends had hallucinated to be the parameters set by God. I will use the construction of Job found in Antonio Negri's *The Labor of Job: The Biblical Text as a Parable of Human Labor* (2009) because of its ontological and ethical reading (the two are the same thing for Negri and, he argues, for Job). In other words, being and ethics are not divided and separated in the story of Job. If this is true then neither is the human and nonhuman divided and separated, for both share some

ELEMENTS OF AN IMMANENTAL ECOLOGY   141

common ontological basis, the same basis that Negri reads into Job (while himself not going as far to the creatural generic as we are): the experience of immense, immeasurable pain. Here the biblical text is not a parable of human labor alone, but of generic creatural labor.

According to Negri's reading of Job, this figure is not pitiful as he stands in pain against a backdrop of tragedy, but is a figure of power as ability or potentiality against Power as constituted and oppressive. In his power Job calls the amoral omnipotence of the divine to account for itself. Such a demand is rhetorically complex, for the protest of Job must not make an appeal to God simply as judge, for *"God is both one of the parties and the judge*. The trial is therefore a fraud."[37] For when Job opens his mouth he will have already condemned himself before the one who judges, as he himself says in the text:

> Though I am innocent, I cannot answer him;
> I must appeal for mercy to my accuser.
> If I summoned him and he answered me,
> I do not believe he would listen to my voice.
> For he crushes me with a tempest,
> And multiplies my wounds without cause;
> He will not let me get my breath
> But fills me with bitterness.
> If it is a contest of strength, he is the strong one!
> If it is a matter of justice, who can summon him?
> Though I am innocent, my own mouth would condemn me;
> Thought I am blameless, he would prove me perverse. (Job 9.15–20
>   NRSV)

By making a defense Job would have to capitulate to the value of justice implicit in the omnipotence of the divine. He would capitulate to an image of value whereby it is just that God, as immeasurable Power, is both the judge and a party to the trial. But in refusing to demand such a trial, in demanding that the omnipotent reveal himself, there is a recognition of the impossibility of a real dialectics in the face of the immeasurable. This parallels precisely the same problem of orthodox theology that sees only death in the struggle of niches as well as in the naturalist who sees "the natural" as the immeasurable Power and source of value.

Negri thinks this relationship and its refusal in the light of the political and philosophical problem of measure:

> *The immeasurable has become disproportion*, imbalance, organic prevalence of God over man. The fact that God is presented as immeasurable demonstrates—once again—that all dialectics are impossible. The trial is not

142   A NON-PHILOSOPHICAL THEORY OF NATURE

dialectical, it is not and cannot be. It is not dialectical because it cannot be "overcome"; or rather, it can be only by negating one of the terms—but this is not dialectics, it is destruction.[38]

Instead Job matches this immeasurable of Power with the immeasurable of his pain. "The book of Job exhibits a sarcastic existentialism that, through pain, denies all dialectics and understands being only as creation. [...] It is necessary to develop power having registered (and dominated) its irreducible passive content and the pain of which power is the daughter."[39] Power in the sense of ability to act or potentiality is the daughter of pain. The creature, as witnessed to most obviously for human beings in the human creature, is able to turn the immeasurability of pain into a source of immeasurable charity and grace. Pain becomes a means of grace, but not a means that comes from the outside of the creature, but an immanent means to the suffering flesh. The immeasurable of charity shares in the immeasurability of pain, for both are that which measures. They become the true measure as immeasurable, as that which can never be measured much like the never-living is beyond the measure of the dialectic of life and death (more on this later). Pain and grace/charity measure the World and reveal that the immeasurability of the world is an immeasurable shame; as a system organized by death and alienation as common (somewhat different from Negri's use of the concept of World and closer to Laruelle's gnostic understanding) the World is but a hallucination of value. Rather, the creation of the World is birthed from the pain and grace of the creatural earth and sea. The World is only absolute as a contingency of creation. Again, consider the words of Negri when he writes:

> But charity cannot be measured because it allows us to participate in the power of creation. In this way the problem of the reconstruction of value can be placed on a new footing. When power opposes Power, it has become divine. It is the *source of life*. It is the *superabundance of charity*. The world can be reconstructed on this basis, and only what is reconstructed in this way will have value; it will continue to not have a measure, because the power that creates has no measure.[40]

Death orders the World because death becomes a common measure to all of life.[41] But in pain this common measure is rendered as simply the object of desire—a desire to eliminate death and pain. To subsume the relative measurable cause of the immeasurable of suffering into a messianic future where the immeasurable of grace reigns. Such grace is the power of production produced by pain. In the story of Job there is a direct correlation between the mismatched dialectical relationship between God and

ELEMENTS OF AN IMMANENTAL ECOLOGY   143

the human being that produces suffering. Job breaks this mismatched dialectic by seeing God. By his protest Job demands that God reveal himself and in so doing Job tears away the absolute transcendence of God. By seeing God, through the immeasurability of God revealed as a body open to vision, Job is able to share in the divine. The immeasurable character of pain and grace is no longer organized hierarchically, but through a simple vision, a knowledge that is salvific. Such vision is creation according to Negri and it is worth quoting his ecstatic hymn to creation at length:

> Job speaks of grace, of the prophecy that anticipates the Messiah. "To see God" is certainly not a moral experience, nor is it merely an intellectual experience. Here the interpreters of the book of Job do what Job's interlocutors, form Eliphaz to Bildad, from Zophar to Elihu, had done: they confine to a given form and measure his experience within the dimensions of the theologically known. And yes, what an incredible experience has unfolded to this point! I have seen God, thus God *is* torn from the absolute transcendence that constitutes the idea of him. God justifies himself, thus God is dead. He saw God, hence Job can speak of him, and he—Job himself—can in turn participate in divinity, in the function of redemption that man constructs within life—the instrument of the death of God that is human constitution and the creation of the world. [...] The antagonism between life and death is resolved in favor of life. My life is the recognition of you—my eyes have seen you. I am. Man is. The backdrop is not modified. It is dominated by the great forces of destruction and death. But man reorganizes himself so as to resist this disease. Creation is the going beyond death. *Creation is the content of the vision of God.* Creation is the meaning of life.[42]

Now, while Negri's ecstatic materialist philosophy (in a way altogether unlike the materialism of most contemporary mainstream Anglophone philosophy) is inspiring, there are certain differences I must mark out. For, from a non-philosophical perspective, Negri tarries dangerously close to affirming the convertibility of life and death, but this antagonism must be seen as formed within the lived. There is no death without the lived, ultimately, but from the perspective of an expanded theory of creatural labor there is a requirement for the recognition of the never-living. The immeasurable of man is indeed pain, but the creation of pain arises out of the relationship between the living, the dead, and the never-living. Thus, whether it is the dialectic between life and death or a nondialectical relationship between the two, there is a third term that stands apart from this relationship and determines it. This is neither God nor the Being of man where the singular meets the universal, but simply the earth as such (and by this I am of course expressing under a more poetic name the biosphere, which

144   A Non-Philosophical Theory of Nature

includes earth, ocean, atmosphere, molten lava, etc.). The never-living
aspect of potential action, the appeal to the earth as immeasurable source
of creation, is what allows for Job to go beyond not just death, but also
the life that births it. For what is it that God appeals to in his justification of himself? In Chapters 38–41, where God makes his justification, he
appeals to creation, including all the living things as well as some fantastic
chimerical monsters that do not exist. These monsters, the behemoth and
the leviathan, are interpreted philosophically by Negri, respectively, as primordial force and the primordial chaos and violence that are the ground
of production, without measure or law.[43] In appealing to his strength, his
Power, God shows Power to be contingent on being able to master this
ground. Interestingly, in the biblical text, while God takes credit for the
creation of both he never comes out and says that he can control them, but
in a rather bombastic style depending on a series of rhetorical questions
merely suggests this.

So what does this ancient biblical story have to do with the contemporary ecological concept of the niche? Negri's retelling of Job is not merely
a parable in a weak sense, but it is an argument concerning the ontological constitution of power as resisting Power. The lived reality of what it
means to be a human subject in pain. We can extend the persona of Job
to creation generally simply by changing some of the terms. So, rather
than Job innocently suffering in the face of a disproportionate and amoral
Power, we have all creatures suffering before a disproportionate and amoral
Power of Nature. Again, I use the uppercase here to signal that I'm writing
about a particular idea of nature, a hallucination of something absolute
in its transcendence to creatures akin to the hallucination of God. This is
Nature hypostasized, the same Nature found in the naturalism of philosophers appealed to for the grounding of ontological and normative claims.[44]
The creation of a niche by a species witnesses to the contingency of such
a Nature. It would not exist without perverse production on earth of new
species. Every time a new species emerges and a niche is formed (remember
immanence is at work here) the suffering of that species calls for Nature
to account for itself. If this cry of violence from the earth and the response
from Nature were to be given in language, what could Nature appeal to in
its justification? For the violence at work in creation is not immeasurable.
It may be overwhelming at times. It may even be evil. But, it is always relative and dependent upon the creation of niches for its existence and in this
way the niche, the creature, is not alienated in its identity by that violence.
By coming into the ecosystem, exchanging energy, it comes to resist and
go beyond death, if only for a moment. The creativity of the niche is the
immeasurable and as such is a certain site of the perversity of nature. Just
as Job was perverse in his acceptance of God's unlimited Power and yet

ELEMENTS OF AN IMMANENTAL ECOLOGY   145

still required that God answer for it, so the niche is perverse in the face of the unlimited Power of Nature.

What is common to creatural being is pain. One species causes pain to another in the working out of niche boundaries. But corollary to this pain is the necessity for biodiversity that niches witness to. There is then a certain creatural sociality as universality at work in the pain of living among one another.[45] This pain is primary and emotions such as fear or anger are but secondary effects that are contingent upon the organization of that pain in the creatural socius. Even violence is secondary to this pain, insofar as that violence can be turned into a peaceable force by way of creation. It isn't my intent to argue for an overturning of death in the ecosystem, but simply to disempower death, just as Job disempowers God. The niche shows that death, as well as life, is secondary to a more immanent creative power at work as nature against Nature. Niches witness to the exile of nature from hypostasized Nature. The refusal of the value of Nature as hallucination of the immeasurable in the name of a grace of nature that is witnessed to in the perverse creative power of new species producing ways of living indifferently to death.

Exchange of Matter and Energy

Up until this point I have only covered what could be referred to as the living and dead elements of ecology as seen through the concept of the ecosystem. At this point we need to transition to those elements normally called abiotic, but which the Chicago-based Irish ecologist and philosopher Liam Heneghan refers to as the "never-living." These never-living elements are outside any dialectic of life and death, a dialectic that operates as a homology where there is a pure confusion of bodies, and this dialectic continues to haunt contemporary philosophies and theologies of nature through their connection to the biopolitical.[46] Because of this homology the ecological conception of the never-living can actually provide a more generic matrix than either neovitalism or nihilistic thought, which is actually based on the anthropocentric notion of extinction, and in so doing may avoid their mistakes by opening up a unified theory of philosophical theology and ecology to a freer, more radically immanent, conception of nature.

In this section I will turn to that element that unifies the living, the dead, and the never-living: the exchange of matter and energy. Energy exchange, also called energy flow, is the productive force at work in the ecosystem and is also constitutive of the ecosystem. Not in a purely unitary way, as if one term were transcendent to the other, or in a way where the two

146   A NON-PHILOSOPHICAL THEORY OF NATURE

terms are simply reversible, but in the sense that the radical identity of the ecosystem and energy are dual. For analysis of the ecosystem and energy seeks "to understand the factors that regulate the pools (quantities) and fluxes (flows) of materials and energy through ecological systems."[47] The terms are not reversible because their identities can be located, even where these identities are always in the midst of the others. Not in a relationship of reciprocity, but simply and radically in-One as the biosphere. Take, for example, the "net ecosystem production," which is the name given to the net accumulation of carbon by an ecosystem.[48] It would be too simple, and all too philosophical and theological, to reduce all the aspects of the ecosystem to being in some sense "carbon." As if all the living and dead elements of an ecosystem could be simply thought of as flows of carbon akin to a computer code. However, this would fail to make any sense of the exchange of carbon, the study of which shows that individuals or actu-alities store and transfer carbon. Without an understanding of the radical identity of these actualities ecologists would be silent in the face of the flux and pools of carbons and yet, against such temptation, ecologists do speak and can trace the carbon balance as it enters the ecosystem through gross primary production (photosynthesis) and leaves through several other pro-cesses.[49] In plain terms, plants are not "merely" carbon flows, because they have processes that are radically separate from the identity of carbon.

We can make the claim, borrowed from Marxism and mutated in Laruelle's non-Marxism, that every exchange is the productive force [*force (de) production*] of work in ecology and still remain scientifically inoffen-sive. Energy is defined rather pragmatically in ecology, drawing on ther-modynamics, as the ability to do work.[50] This definition of energy was first adapted for scientific ecology by Howard Thomas Odum, brother of Eugene Odum discussed earlier. Howard Thomas Odum used the cyber-netic theories of the 1950s alongside of the general laws related to energy derived from research in theoretical physics to develop theories about and research projects concerning energy flow.[51] This definition of energy is given alongside the two laws of thermodynamics by Odum:

> Energy is defined as the ability to do work. The behavior of energy is described by the following laws. The *first law of thermodynamics* states that energy may be transformed from one type into another but is never created or destroyed...The *second law of thermodynamics* may be stated in several ways, including the following: No process involving an energy transforma-tion will spontaneously occur unless there is a degradation of the energy from a concentrated form into a dispersed form.[52]

This pragmatic definition of energy allows the ecologist to locate the immanent process of energy in food webs, the productivity of species, the

ELEMENTS OF AN IMMANENTAL ECOLOGY   147

various cycles of biogeochemical elements such as carbon and nitrogen, and the other aspects of the ecosystem, without overdetermining the identities of these elements *as* energy.

Without reducing ecology simply to non-philosophy, there is nevertheless some validation of Laruelle's conception of non-philosophy's scientific character for in ecology there remains a unilateral duality at work. There is energy/matter and there are the various organisms that are in process through the energy flows, but ultimately there is a weak autonomy of the organism in relation to energy and matter and a radical autonomy of energy. This can be seen more clearly when you consider that all energy exchange takes place within some system or organization that is either open or closed. Lévêque summarizes the system of energy exchange across the biosphere (or earth) this way: "The earth is an open thermodynamic system: the flow of solar energy (high energy photons) that penetrate the biosphere are gradually transformed into work and heat, which is ultimately dissipated into space in the form of infrared radiation (low energy photons)."[53] Chapin, Matson, and Mooney's description helps round this out when they write, "Most ecosystems gain energy from the sun and materials from the air or rocks, transfer these among components within the ecosystem, then release energy and materials to the environment."[54]

In other words, the biosphere is an open system that receives energy anew each day from the sun, which is stored in living organisms and released through various processes, including death. The identity of the organisms is then unilaterally determined by energy and material in a sense that avoids the necessity of a philosophical materialism or reduction. For, at the start of the energy cycle is the radiation from the sun. This is captured by plants, which in turn transfer this energy to animals as well as microorganisms. In each case the organism is dependent at some point in time on a single flux of energy that takes a different form every time as it is transformed and turned into work and heat by abiotic elements such as water, atmosphere, and soil materials (which are themselves derived from rocks that have been broken down).[55] As Lévêque states, "The emphasis is on the notion of flux and no longer only on that of mass."[56] Thus the ecosystem is an open flux, insofar as it both derives its energy from outside the system and energy is lost from the system.[57] This openness can be described as a "constant dissipation of energy" that takes place in the food webs where energy is turned into material by decomposers and then used by autotrophs (organisms that extract raw materials from the mineral world and synthesize all the molecules needed for their functioning—plants are a common example).[58]

How can this energy be understood from the perspective of an ecosystem (of) thought? In the introduction to this chapter I made the claim that

148   A Non-Philosophical Theory of Nature

thought can be treated like a real ecosystem and as part of a real ecosystem. From an ecological perspective thought is also dependent upon this single source of energy. For thought to do work it requires energy. A simple and likely obvious statement, but one that is rarely considered when doing philosophical or theological work. But do thoughts decompose? Do they live and die, passing on their energy to something else? Let us hazard a yes and let this yes be said without an appeal to crude scientism or reductive materialism. But this yes must also be said without an appeal to pseudo-science or narcissistic notions that our thought creates reality. What then would it mean? An enigmatic answer: thought transfers its energy in the everyday. What this means is that thought is an ecological process just as respiration or the carbon cycle is. The function of this "just" needs to be clarified. For the point isn't that thought is "just material" as if we knew that the value of material was low from the perspective of human freedom. Rather, the point is that thought and the rest of the ecosystem has a certain equivalency in terms of their relative autonomy in relation to the radical autonomy of the never-living. Thought isn't something to deride, but neither is it something divine that can only secure its divinity by debasing what is taken to be beneath it. Thought carries with it energy that can be transferred to other forms of thought, can animate wider ecosystems (of) thought, or can be used to direct action. But in this way thought remains relative, it has no absolute autonomy, but is unilaterally determined, at least on this refractory planet, by the sun and so any animation or action is always as a custodian rather than a master.

So this is not a metaphor. I am claiming that within thought there is energy that is transferred among living thought, dead thought, and the never-living of thought. The definition of energy inherited from general thermodynamics guides us here: energy for thought is the ability to do work. Thus we don't need to make any appeals to an already existing materialist philosophy or to a reductionist account of thinking that says thought "just is" energy. Such claims are often incoherent for they are dependent upon a circle of meaning that they claim to escape. For many such claims are often pitched against the reenchantment of the World and in favor of a further disenchantment. Consider the words of Ray Brassier who claims that

> the disenchantment of the world understood as a consequence of the process whereby the Enlightenment shattered the "great chain of being" and defaced the "book of the world" is a necessary consequence of the coruscating potency of reason, and hence an invigorating vector of intellectual discovery, rather than a calamitous diminishment. [...] The disenchantment of the world deserves to be celebrated as an achievement of intellectual maturity, not bewailed as debilitating impoverishment.[59]

ELEMENTS OF AN IMMANENTAL ECOLOGY   149

So, reducing the World to a set of "just is" propositions is supposed to bear
witness to a sign of "intellectual maturity" for they show that the thinker
is involved in thinking the truth regardless of what that truth may be. The
consequence of this rhetoric is obviously intended to turn the reductionist
into a kind of hero of truth, but surely the hero *just is* valorizing himself
within a circle of meaning it hasn't been able to reduce in its practice.

Instead of playing such competing games of meaning, why not think as
the wave of energy. Disinterested in the World, whether it is enchanted or
not, but acting nonetheless. For thought clearly begins with energy. When
I sit to think about an ecosystem (of) thought my body is burning calories.
My body that is thinking requires energy; it requires that I eat if I am to
think. It requires too that I have eaten well, that I have eaten nutritious
food throughout the living action of my body. For nutrition is necessary
for my brain to form synapses correctly and allow for synaptic connections. These connections and the thought that is thought are immanent
to one another. For while these connections allow for the production of
thought, they indeed are the production of thought, they do not occur
within a vacuum. For when I think I do so with the ideas of others, with
common notions as well as with presuppositions. These play a part in the
form of whatever thought I go on to think. For example, when I sit down
to read Spinoza's *Ethics* there is a transfer of thoughtful energy from the
dead thought on the page to the living thought transforming the material of the dead thought in the midst of my thinking it. Writing is but a
material trace of the living thought that provides the material energizing
the living thought. Ideas thus exhibit an energy-like quality. When I read
the ideas of some thinker I am able to do work with my own thought. My
own thought is able to perhaps do the same for another body of thinking.
The reality of this image of thought (seen from the vision-in-One) is more
adequate to what actually happens when we think. For we do not think
with perfect understanding, simply repeating the master, but always akin
to taking some material from one world to another. Sometimes that matter
is carried on our shoe, sometimes in a new configuration of our body. But
this is how we actually think in the moment: contaminated, energized,
and without the transcendent master.

Space and Time

The exchange of matter and energy is an important never-living aspect
of ecological theory. But ecosystem structure and functioning are governed by at least five independent control variables called "state factors":

150   A Non-Philosophical Theory of Nature

"climate, parent material (i.e., the rocks that give rise to soil), topography, potential biota (i.e., the organisms present in the region that could potentially occupy a site), and time."[60] Of these we have effectively covered potential biota (biodiversity), parent material (material exchange), and to some extent climate insofar as climate is closely linked with the energy flow. Space and time are important elements of the ecosystem process, but how they are important can be illuminating of the general structure of ecosystems (of) thought generally as well as questioning the commonsense views of time and space.

There is no real need for technical language in this section, for the effects of space on ecosystems are relatively obvious. Ecosystems are partially determined by where they are on the planet in relation to where the sun is. That climate may also be determined in part by landform effects. For example, mountain ranges affect local climate through what are called "orographic effects," simply referring to the presence of the mountain. One example of this is what is called a "rain shadow," or zone of low precipitation downwind of the mountains caused by the cooling and condensation of air moving up the windward side of the mountains. Deserts and steppes often exist because of their downwind proximity to a mountain range.[61]

Time functions in a similar way and Chapin, Matson, and Mooney summarize the importance of time in ecosystem processes this way: "Time influences the development of soil and the evolution of organism over long time scales. Time also incorporates the influences on ecosystem processes of past disturbances and environmental change over a wide range of time scales."[62] For organisms too there is a temporal element in their ecosystem functioning that is important if one is to understand the overall ecological processes at work: "Many species may change their trophic level [i.e., where they can be placed in terms of the exchange of matter and energy] during their life cycle. A fish may be planktivorous in the larval stage, a consumer of invertebrates in the juvenile stage, and piscivorous in the adult stage."[63] Ecological processes take time and in time there is a certain heterogeneous character to the ecosystem and its elements that can be witnessed.

What is common to both space and time in ecological theory is their heterogeneity. Lévêque goes so far as to claim that there is a "consensus" in ecology "that nature is heterogeneous."[64] He goes on to define spatial heterogeneity as existing from either a static or dynamic point of view. "An environment is heterogeneous if a qualitative or quantitative variable, such as plant cover or air temperature, has different values in different places. However, in functional terms, heterogeneity is also apparent when there is a change in the intensity of functional processes in response to variations in the structure of the environment."[65] He goes on to define temporal heterogeneity writing that "[t]emporal heterogeneity, more often

ELEMENTS OF AN IMMANENTAL ECOLOGY   151

called temporal variability, can be defined simply by the different values taken by a variable in a single point of space as a function of time."[66] The two dimensions are closely related, though, such that "structural heterogeneity can be defined as the complexity and variability of a property of an ecosystem in time and space."[67]

This underlying heterogeneity is an affront to a popular strand of environmental philosophy that has arisen in response to the practice of ecological restoration. Ecological restoration is often faulted for a confusion at the temporal level for the restorationist aims to restore some ecosystem to some prior state. This prior state is often chosen on the basis of something somewhat arbitrary. For instance, in the context of the American Midwest prairie, there are attempts to restore the prairie ecosystem simply because of the rarity of the prairie at the global level. The ecosystem that replaced the prairie functions, but it is also rather homogenous and common throughout the neo-Europes (a name Crosby gives to those ecosystems that have come to closely resemble the ecosystem structure of Europe linking European social imperialism with an ecological imperialism as well).[68] Yet for Eric Katz, one of the harsh philosophical critics of ecological restoration, there is an underlying problem here of authenticity. Katz writes,

> A "restored" nature is an artefact created to meet human satisfactions and interests. Thus, on the most fundamental level, it is an unrecognized manifestation of the insidious dream of the human domination of nature. Once and for all, humanity will demonstrate its mastery of nature by "restoring" and repairing the degraded ecosystems of the biosphere. Cloaked in an environmental consciousness, human power will reign supreme.[69]

He goes on to compare the restored ecosystem to a piece of forged art as if the ecosystem were analogous to an artistic creation.[70] Yet he passes over so-called artifacts made by nonhuman animals, like a beaver dam, suggesting that only human beings are able to create artifacts and exist outside nature.[71] But a philosophical notion like authenticity isn't able to understand the ecosystem because the identity of its space and time simply is *heterogeneity*.

Resilience

Resilience is a recent ecological concept that is important to ecological management, of which ecological restoration is a part. The notion of management is somewhat tainted, especially for radical thinkers, by the association it has with techno-capitalist biopower. However, management in an eco-logical context, while it may refer to certain kinds of nefarious biopolitical aspects of the human-social relationship with the wider biosphere, tends to refer simply to a more intentional and scientifically grounded relation-ship with the natural world. Resilience thinking begins first by rejecting any kind of dualism between humans and nature, because human beings are a part of ecological processes: "We are all part of linked systems of humans and nature" which are referred to as "social-ecological systems."[72] This notion of a social-ecological system is important for dissolving false problems common to environmental thought, specifically with regard to the notion of "sustainability." Radkau's history of social-ecological systems shows that most environmental management decisions tend to be made with a poor understanding of the system in general. Walker and Salt sug-gest that this poor decision comes from focusing on "isolated components of the system."[73] Such an approach is common to the economic approach to ecological management with its focus on "efficiency" and "optimiza-tion." But in a dynamic system like the biosphere there is no simple "opti-mal" state as present in economic reasoning where one simply needs to be more efficient with resources.[74]

What Walker and Salt show through five case studies is that most attempts to optimize simple elements of a complex social-ecological sys-tem lead to a diminishing of that system's resilience.[75] Walker and Salt put this starkly writing, "A drive for an efficient optimal state outcome has the effect of making the total system more vulnerable to shocks and disturbances."[76] Against this economically modeled idea of optimization resilience thinking begins from the simple idea that "things change" and "to ignore or resist this change is to increase our vulnerability and forego emerging opportunities. In so doing, we limit our options."[77] Change is in and of itself value neutral because "[t]here is no such thing as an optimal state of a dynamic system. The systems in which we live are always shift-ing, always changing, and in so doing they maintain their resilience—their ability to withstand shocks and to keep delivering what we want."[78]

So resilience focuses on two central research themes: thresholds and adaptive cycles. These two themes underpin resilience thinking and it is worth quoting Walker and Salt's definitions at length because of their clarity:

Thresholds: Social-ecological systems can exist in more than one kind of stable state. If a system changes too much it crosses a threshold and beings behaving in a different way, with different feedbacks between its compo-nent parts and a different structure. It is said to have undergone a "regime shift." [. . .]

ELEMENTS OF AN IMMANENTAL ECOLOGY   153

Adaptive cycles: The other central theme to a resilience approach is how social-ecological systems change over time—systems dynamics. Social-ecological systems are always changing. A useful way to think about this is to conceive of the system moving through four phases: rapid growth, conservation, release, and reorganization—usually, but not always, in that sequence. This is known as the adaptive cycle and these cycles operate over many different scales of time and space. The manner in which they are linked across scales is crucially important for the dynamics of the whole set.[79]

The rest of this section will focus on filling out these concepts.

Walker and Salt illustrate the thresholds theme with a simple analogy. Imagine that the system is a ball and that ball is placed within a basin that indexes the system's "state" variables. A simple system could be simply the number of fish and the number of fishers, but it is important to keep in mind that a system is of course fractal and is thus n-dimensional. Now within this basin the ball will tend to roll toward the bottom, that is, in system's terms, it tends toward equilibrium. But equilibrium is not static and is rather always changing based on the changing external conditions and so the shape of the basin is itself always changing as the conditions themselves change. The shape of the basin can change so much, meaning the conditions can change so much, that the ball can suddenly find itself rolling into a new basin with a different fundamental organization. To put some flesh on this example, say a particular linchpin species has a small niche width due to some other external change. If that species continues to be pushed out of the ecosystem functioning toward extinction, then the system will fundamentally change. While prior to this change you could adjust on the basis of that species, that possibility is no longer open to you because the structure of the basin has fundamentally changed.[80] Walker and Salt summarize this thus: "Once a threshold has been crossed it is usually difficult (in some cases impossible) to cross back."[81] Furthermore, sustainability requires that we learn if and where thresholds exist and figure out how to manage the capacity of the system in relation to those thresholds.[82]

Adaptive cycles give some organization to the changes that occur with ecosystems. "One important aspect about cycles is recognizing that things happen in different ways according to the phase of the cycle the system happens to be in."[83] There are four phases to the adaptive cycle: the rapid growth phase (or r phase), the conservation phase (or K phase), the release phase (or Ω phase), and the reorganization phase (or α phase). The r phase is the early period of the system where there is rapid growth "as species or people [...] exploit new opportunities and available resources."[84] During this phase species are exploiting every possible niche and the system is

154   A Non-Philosophical Theory of Nature

weakly interconnected and has a weakly regulated internal state. The K phase occurs when connections between different actors increase and all competitive advantage moves to those actors who can "reduce the impact of variability through their own mutually reinforcing relationships."[85] During this phase the system becomes more and more rigid, which increases efficiency but results in a loss of flexibility. The Ω phase occurs when some shock occurs to the rigid K phase. Walker and Salt claim that "[t]he longer the conservation phase persists the smaller the shock needed to end it."[86] This results in a certain ecological form of "creative destruction" as all the resources that were tightly bound within their relationships are suddenly free and released to new relationships. This phase leads quickly to the α phase where "[s]mall, chance events have the opportunity to powerfully shape the future. Invention, experimentation, and reassortment are the order of the day."[87] During the Ω phase there is no stable equilibrium, the counters of the basin are not operative. "The reorganization phase begins to sort out the players and to constrain the dynamics. The end of the reorganization phase and the beginning of the rapid growth phase is marked by the appearance of a new attractor, a new 'identity.'"[88]

Now Walker and Salt also tell us that this description of the adaptive cycle does not necessarily lead to resilient systems that we would want to encourage. In fact, the adaptive cycle could end in the ecological equivalent of a poverty trap if the cycle were to happen within a basin that already has constrained options through homogeneity. Moreover, though this description of the cycle is generally the way a system passes through the phases, there are other open possibilities save for the system going directly from a release phase to a conservation phase.[89]

What is the relation between identity and the phases at work in resilience thinking? Is the identity of the ecosystem what happens only in the conservation phase, when it is most rigid? Or is it the moment when it all falls apart during the release phase? What is the relation of the constant change at work in the ecosystem to identity? Is there a priority of difference here? Such a priority is dubious and introduces the philosophical and reversible dyad of identity and difference into ecology. For if there were a priority of difference over identity at work in resilience thinking, how would the ecologist study the phases as such? The answer is that the identity of the ecosystem is separated from the phases that it manifests or that are "cloned" from the ecosystem. We can call the phases a clone of the ecosystem because at its radical immanence the ecosystem is simple and inconsistent (meaning diverse in act). For the ecosystem is the exchange of the material and energy from the dead and the never-living to the living. That remains regardless of the phase the ecosystem manifests. Even when there is no stable equilibrium or if the equilibrium is rigid the ecosystem

ELEMENTS OF AN IMMANENTAL ECOLOGY 155

nevertheless = One. What resilience thinking then shows is that, while there is no equilibrium or balance in nature, there is a certain organization to the changing dynamics of an ecosystem. This gives the ecosystem an identity without requiring that this identity be anything like a stable, harmonious whole. Rather the standard philosophical and theological notions of identity are challenged here and it gives rise to something altogether more in line with the non-philosophical conception of identity. Again, this isn't to shore up non-philosophy's concepts, but to show where non-philosophy has already been mutating standard philosophical conceptions by way of a scientific posture.

Chapter 10

Ecologies without Nature

In the previous section we traced six principles and concepts from ecology that, when unified with philosophical theology, can change our understanding of nature and help us to furnish a new theory of nature. But before we offer that theory of nature we must first deal with a two potential challenges to the project of a theory of nature formed from a unified theory of philosophical theology and ecology. For there is another option for philosophers than this notion of nature given from the impossibility of foundation: one may simply reject nature altogether. Timothy Morton and Bruno Latour have offered the strongest philosophical arguments against the validity of the concept of nature in relation to ecology and so their criticisms must be responded to. Ultimately I will argue that while ecology as a science may not require the philosophical concept of nature to function, nature is still a "good name" (as Derrida said of God). The reason nature remains a good name is, in part, because of its ability to confound philosophical thinking, but there is always the risk that this confusion be wielded as a weapon against human beings and other creatures. Morton and Latour, in similar ways, attempt to decommission the weaponized form of Nature. Nature is written here with the capital N to mean a metaphysical conception of nature, as transcendent in philosophy, but I will keep the use of the alternating use of a capital N and lowercase n to a minimum. So, my response to Morton's and Latour's criticisms will not be an outright rejection but, in true non-philosophical style, an attempt to radicalize their criticisms, to make them immanent, to a conception of nature *for* creatures but *against* the World created by those creatures.

Latour's work will be the starting point, as it is, despite Latour's prominence, the weaker argument of the two. The weakness of that argument will be developed in relation to Morton's own argument, because Morton,

158   A NON-PHILOSOPHICAL THEORY OF NATURE

as we will see, attempts to think thought itself as ecological by way of scientific ecology, whereas Latour attempts to think ecology merely politically by really only considering political ecology.

Latour provides for us a succinct definition that encapsulates the philosophico-political understanding of nature that he is writing against:

> Nature: Understood here not as multiple realities [...] but as an unjustified process of unification of public life and of distribution of the capacities of speech and representation in such a way as to make political assembly and the convening of the collective in a Republic impossible. I am combating three forms of nature here: the "cold and hard" nature of the primary qualities, the "warm and green" nature of *Naturpolitik*, and finally the "red and bloody" nature of political economics. *To naturalize* means not simply that one is unduly extending the reign of Science to other domains, but that one is paralyzing politics. Naturalization can thus be carried out on the basis of society, morality, and so on. Once the collective has been assembled, there is no longer any reason, by contrast, to deprive oneself of expressions of common sense and to use the term "natural" for something that goes without saying or something that is a full-fledged member of the collective.[1]

The entirety of Latour's argument is sketched out in this short definition of nature. First, he defines nature in a particularly univocal way as that which finds expression in three other modes of nature. This understanding of nature is tied up intimately with a certain understanding of Science, which when capitalized refers to "the politicization of the sciences" and is opposed to *the sciences*, which are vital to the Republic in their plurality.[2] Against this vision of Nature and Science (again with the capitals intended) Latour poses the Republic. However, Latour's conception of the Republic is not purely human, but instead refers to the collective of "public things," human and nonhuman, that are given a certain "due process" under a new form of Constitution.[3]

Of course these two concepts do not carry their usual meaning either. For due process is subtracted from its purely legal meaning and extended to nonhuman things as well. Here the concept of the Republic really determines the meaning of due process, because Latour plays on the underlying Latin meanings of the two roots in Republic—*res publica* or "public things."[4] In Latour's recasting of political ecology without nature he conceives of a new form of government that recognizes the ecological connections between things, connections that require that we give "due process" to understanding nonhuman things from a democratic perspective. Due process is then simply "taking the time" to understand those relations, with all the difficulties of debates and attempts at persuasion that come with parliamentary democracy.[5]

ECOLOGIES WITHOUT NATURE   159

This form of democracy too must be changed by way of political ecology without nature; for the current "constitution" was based on a two-house collective that posited a single "Nature" and multiple societies. These two houses are taken to be political assemblies: one an assembly of things and the other an assembly of humans.[6] Latour makes an interesting insight into this model of politics, pointing out that it is based on a double split. The obvious split is between nature and society, but there is also a split within society between those who are supposed to know, the "Philosopher-Scientist" who is able to leave "the Cave" and then bring back knowledge to those who remain in the Cave, those who practice science, and those who then decide what actions to perform on the world of nature represented by the Philosopher-Scientist.[7] In distinction to this regime the Republic of the collective removes the determination of Nature and society and places them relative to a single Republic (keeping in mind Latour's particular meaning) already existing out of the Cave that is constantly in the process of exploration or due process.

So, why is Latour rejecting nature here? "Because nature is not a particular sphere of reality but the result of a political division, of a Constitution that separates what is objective and indisputable from what is subjective and disputable."[8] His construction of an alternative "Constitution" is often strong and his own project marks a kind of "democracy of thought" that would seem to agree with the non-philosophical vision. The differences between Laruelle and Latour are, despite all of this, rather stark. A full elaboration of those differences is outside the remit of this work, albeit perhaps needed, as both are major French thinkers interested in bringing science and philosophy together in radical ways at a time when that position was unpopular in radical philosophy circles. Rather I am going to focus on a criticism of Latour not directly located to the difference between his work and that of Laruelle, but related instead to the overdetermination of his thought by "the Political." I am following Latour here in capitalizing the Political to refer to his metaphysicalization of politics that determines his conception of *political* ecology without nature. For the problem with Latour's "thoughtful democracy" (in opposition to a democracy (of) thought) is that it is overdetermined by way of parliamentarianism. Latour casts this idea as original, as if a "parliament of things" marks a real turning point in the relationship between humans and nonhuman things. Yet, though less intentionally developed than Latour's "new Constitution," we do find traces of this open, nonmodern approach in the past of environmental history, where societies did not have a "double split" between nature/culture and those-who-know/those-who-do. As Joachim Radkau has argued in his *Nature and Power: A Global History of the Environment* (2008) the various forms of this already existing tacit

160   A NON-PHILOSOPHICAL THEORY OF NATURE

ecological knowledge has a mixed record in terms of fostering sustainable human/nonhuman-coupled ecosystems.[9]

If such a Latourian Republic, governed by a parliament of things, is already at question in those forms that have already existed, then it is wholly *philosophical* in genre and thus specular rather than a form of thought that escapes philosophical boundaries. In short, Latour's own interesting conception of political ecology still practices philosophical subsumption by requiring that ecology subsume itself within what he calls a "political epistemology." While his criticism of Nature is instructive, it is instructive from a philosophical rather than ecological perspective. Latour's political ecology without nature is also, in terms of the structure of its thought, *without ecology*. It does not allow the scientific practice of ecology to challenge the philosophical "parliament of thought" he proposes. Furthermore, while we may have left the prison of the Cave, Latour still requires a policing and one that may be more insidious and more intense. By casting lived reality in the mold of a parliament Latour binds thought to another authority, even while he claims he is opening up both philosophy and the sciences to the greater reality beyond the usual authorities of thought. But he does so without challenging the underlying authority of capital itself.

In distinction to this Timothy Morton identifies capitalism in a way that brings to mind Deleuze and Guattari's "disjunctive synthesis" as a negative form of ecology. He writes, "Capitalism has brought all life forms together, if only in the negative. The ground under our feet is being changed forever, along with the water and air."[10] Morton's investigation is more explicitly historical in its analysis of nature and so he is able to show how the conception of nature operative in romantic environmentalism is connected intimately to capitalism.[11] By not confronting the reality of capitalism Latour allows it to remain in the background of the Republic becoming a new kind of "environment," a new kind of nature about which one can only say "it's like that." By this "it's like that" I am thinking here of what Christian Jambet and Guy Lardreau portmanteau of the phrase *c'est comme ça* as the *sékommça* operative in political naturalism, which will become more important in part IV.[12] Latour's argument against nature falls short on a number of levels then. First, it fails to address capitalism, which like nature also is an "unjustified process of unification of public life," and so even if it succeeds in creating a political ecology without nature, it does so by creating a political ecology embedded within capitalism. Second, despite Latour's work in science studies, the form of ecology worked with is always "political ecology." The sciences are abstracted so far away from the authority of Science that scientific ecology is given no particular ability to challenge philosophical thought, even as the political model of parliamentarianism is.

ECOLOGIES WITHOUT NATURE   161

This being the case Latour's argument against the idea of nature does not really touch on the specific question at issue here—that of the relationship between the idea of nature and ecology—even if he is right to criticize the standard philosophical form of nature. Morton's argument though is concerned more explicitly with a kind of "unified ecology" that he calls "the ecological thought."[13] His theorization of the ecological thought, which he says "is a virus that infects all other areas of thinking," engages specifically with concepts from scientific ecology while still arguing for a kind of political ecology at the same time.[14] So, his attempt to cast an ecology without nature must then bear on our attempt to recast nature non-philosophically and thus requires a response.

What does Morton mean by nature? Morton shares Latour's criticism of modernity as the site where the representatives of Science (as opposed to the sciences) rely on a certain idea of Nature to stop thinking: "[I]n general the scientisms of current ideology owe less to intrinsically sceptical scientific practice, and more to ideas of *nature*, which set people's hearts beating and stop the thinking process, the one of saying 'no' to what you just came up with."[15] Nature then operates in a similar way to the usual philosophical transcendental authorities that are treated as simple material in non-philosophy, and there is much to be sympathetic with from a non-philosophical perspective when Morton identifies nature as that transcendental object of nature that takes the environment and turns it into a fetish object.[16] Or, perhaps especially, when Morton identifies (though with little elaboration) that there is a deep connection between environmental thinkers such as Aldo Leopold, who certainly remains beholden to a transcendental idea of nature as underlying structure and often positing a divide between humans and nature (however *minimal* it may be), and Gilles Deleuze and Félix Guattari's "new and improved" conception of nature.[17]

This criticism remains philosophical and so not entirely a challenge to our proposed rethinking of nature non-philosophically arising from an immanental ecology (or unified theory of ecology and philosophy). Yet, Morton goes beyond Latour because he tries to conceive of *ecology* without nature, or rather takes it that, in its actual practice, ecology already operates without the philosophical conception of nature: "Why 'ecology without nature'? 'Nature' fails to serve ecology well. I shall sometimes use a capital *N* to highlight its 'unnatural' qualities, namely (but not limited to), hierarchy, authority, harmony, purity, neutrality, and mystery. Ecology can do without a concept of something, a thing of some kind, 'over yonder,' called Nature."[18] In short, with regard to ecology and the ecological thought that spills over science to infect all forms of contemporary thinking, "nature" is a bad name. It is a bad name for Morton precisely

162   A NON-PHILOSOPHICAL THEORY OF NATURE

because "nature is often wheeled out to adjudicate between what is fleeting and what is substantial and permanent. Nature smoothes over uneven history, making its struggles and sufferings illegible."[19] In this way the idea of nature covers over the truth of the flux of living beings and provides a kind of anchor of meaning that the monotheistic God once provided—nature is a kind of Big Other in the Lacanian sense.[20]

At a deeper level, though Morton does not explicitly identify this, nature is a bad name because it isn't infected by the ecological thought. In a sense Morton leaves nature forsaken by the ecological thought; as a secularized form of God, nature may be crucified in Morton's mesh, but it is not resurrected and as such it is only murdered. This quasi-theological description is not an unduly foreign element being forced on Morton's ideas, but rather exploits his own aligning of his ecology without nature with Derrida's attempt to radicalize negative theology's "without" [sans]: "Derrida's profound thinking on the 'without,' the sans, in his writing on negative theology comes to mind. Deconstruction goes beyond just saying that something exists, even in a 'hyperessential' way beyond being. And it goes beyond saying that things do exist. 'Ecology without nature' is a relentless questioning of essence, rather than some special new thing."[21] This quasi-theological negation of nature will be important to our own project, but insofar as Morton uses this negation to cut off nature from redemption, it is only half-theological or is not radically theological. Will Morton's proclamation of the "death of nature" not require a madman to proclaim that this death was really a murder and to ask how we are to become worthy of it?

Morton's casting of nature as forsaken suggests an ethical failure on the very terms laid out in *The Ecological Thought* (2010). For, in terms we will make clearer later in this chapter, while there is certainly extinction within thought, nature is not yet dead. Even if, within scientific and philosophical discourses, the essentialist conception of nature is finally passing away, Morton rightly locates a number of other forms of nature (which he mockingly calls "new and improved") that are still operative in thought. Morton's own interesting discussion of the ethical basis for an ecological awareness without nature suggests that we are responsible for the biosphere "simply because we are sentient. No more elaborate reason is required."[22] As a kind of ethical form of gnosis Morton's suggestion is powerful. At the level of lived immanence there is no formal creation of ethical responsibility, as if structured by law, but the coexistence with others simply requires care, regardless of our own personal place in terms of causation. If this is the case, ideas too require care, as they too are something damaged or at risk: "One of the things that modern society has damaged, along with ecosystems and

ECOLOGIES WITHOUT NATURE 163

species and the global climate, is thinking."[23] By declaring nature as such a bad name Morton is advocating a kind of "cowboy ecology" of thought. He is actively treating the idea of nature as a varmint, suggesting that it be actively destroyed without considering its place in the ecosystem or its resilience as a population (of) thought active in that ecosystem.[24]

Nature is still important for thinking, specifically philosophical and theological thinking, and especially thinking in light of the environmental crisis. But it is important to rethink nature from within "the mesh" and the ecological thought or, in our terms, from an immanental ecology. From this perspective it isn't that ecology needs nature or that ecology should think from nature, but that nature needs ecology even as nature is a kind of "under-determination" of ecology. So ecology without nature, by all means, but by way of nature with ecology.

Conclusion

What is the status of the old divisions of thought after the work carried out over the past three chapters? Theology has still not been returned as the queen of the sciences. Science can still operate with relative ease without the need of worrying about the transcendental conditions for their thought. Philosophy still is but one way of thinking alongside a plurality. In a certain way immanental ecology does not undermine the reality of the divisions of thought at all, for these divisions were always contingent forms of the organization of thinking. Immanental ecology, as a form of the non-philosophical practice of theory, simply allows us to see that the old divisions of thought were never true divisions. That the relative identities we can bestow on philosophy or theology is always a matter of tendency, a tendency in the light of the radical identity of the Real-One. In-the-last-instance ethics, theology, philosophy, ontology, God, man, creatures, and all of creation are One. Their identity is but an effect of the radical immanence of the One, as the radical immanence of the lived. These divisions of thought are only useful when they are useful. This tautologous statement is not meaningless, but rather it points to the material basis of thinking about thought. If a division of thought is productive of thought, then divide, but if the division blocks thinking, then step back from your fabulation and refabulate from the experience of thinking itself rather than from a hallucination of its image.

It is worth noting again Laruelle's notion of "under-determination" in order to understand the status of these different regions of thought or, as

164   A NON-PHILOSOPHICAL THEORY OF NATURE

they take themselves to be, these different regimes of thought. Laruelle
writes:

> The fusion "under (captial-S) science" [*sous (la) science*] indicates an
> inequality of forces and status, the "under" no longer means a domina-
> tion or even an overdetermination like in the works of Marx concerning
> the relations of production, but an immanent under-determination of the
> philosophical subject by science, no longer a return of science on itself with
> the risk of specularity, but a superposition marked by a line of definitive
> contingency.[25]

In more familiar terminology, the vision of the division of science and phi-
losophy/theology at play here is one where the two unequal forces of thinking are brought together into a single, new thought. The formalism for
superposition or idempotence can be expressed as $1 + 1 = 1$. For Laruelle
the different forces of thoughts are like quantum waves that can meet and
form, not a combination of the two waves, but an actual new wave where
there are three distinct identities now. We're suggesting a similar image of
thought where the forces of thought are being treated as ecosystems. The
elements of these ecosystems are brought together, not in a combination,
but as a new ecosystem of thinking.

In part IV, I will propose a theory of nature from the perspective of a
unified theory of philosophical theology and ecology that will begin to use
this immanental ecology. I will engage with a number of different ecologies
(of) thought, separating out species and weakening their self-sufficiency by
placing them within wider spatial and temporal scales. The desire is that
by doing this we will not only treat these ecosystems (of) thought as simple
material, but will do so as ecological material with the intention of deriving
a theory of nature that is immanent to that material and can be adequately
called a first name for the Real and finally what such an adequacy might
mean. While the next section arises out of this immanental ecology it does
so insofar as I am now able to treat philosophical and theological conceptions of nature as an ecological material. What this means is that I
can separate out the different uses of the concept of nature as manifest-
ing a particular functional form of nature as ultimately perverse. Thus
for Aquinas nature is presented as creation or creatural, for Spinoza there
is a chimera of God or Nature, and for Abu Ya'qûb al-Sijistânî and Naṣîr
al-Dîn Ṭūsī there is recognition of the fundamental Oneness of nature
prior to Being or Alterity.

Part IV

A Theory of Nature

Chapter 11

Separating Nature from the World

Introduction

In this final part I will turn my attention to developing a theory of nature using the methods of non-philosophy in the form I have given it in this work, that is, as a unified theory of philosophical theology and ecology. This theory of nature will be pursued and developed by way of philosophical and theological material "under-determined" by the immanental ecology sketched out in the last chapter. I follow ecology here insofar as I am able to avoid the need to talk about some hypostasized thing called "Nature." Rather nature here comes to be a particular name that is productive of thinking or, as Laruelle would say, a force (of) thought. This productivity would be an effect of a conception of nature that is but a name. A non-thetic transcendence that is but a manifestation, as a first name, of the radical immanence of the Real. The point is not, then, a theory of nature within a kind of naturalism developed from ecology rather than from empiricism, but a form of nature that breaks with these sorts of philosophical and theological humiliations of the creatural and instead becomes something of use to creatures. As we said in the last chapter, ecology without nature, by all means, but by way of nature with ecology. By way of a theory of nature that breaks with the circle of Narcissus and Echo present in philosophy and theology by introducing the posture of scientific ecology into philosophical theology. Such a break, which begins with the perversity of nature, aims to think nature as irreducible to an idealized matter (i.e., materialism) nor reducible to a transcendental idea that forms matter (i.e., naturalism).

The chapter begins by explaining in more detail the approach to the history of philosophy and theology present here. It then turns to the dominance of the metaphysical concept of World that is present in most philosophies of nature. I illustrate this dominance by way first of Heidegger and then of Badiou; two figures chosen for their relative distance from one another and in the hope that this distance yet structural homology with regard to nature will illustrate what I take to be an invariant worldliness at work. I then turn again to Heidegger and his conception of the fourfold, because of Heidegger's great influence on environmental thought and philosophy and theology more generally. Thus his thought will work as a kind of index for those thinkers who build off this same metaphysical dominance of World. I take the fourfold as a sketch of a formalist conception of the World taken to be nature in philosophies of nature, where nature is but a thing that is known ultimately through various relations it has within a World. I then unilateralize this fourfold in order to derive a different identity of nature beyond the being of a thing. This unilateralization is then explained further by way of a creative ecological recasting of the debate between Thomistic ecosystems (of) thought (i.e., theologies of analogy) and Spinozist ones (i.e., philosophies of immanence). This will bring to the fore the constant connection in the history of thought between thinking God and Nature. Finally, I will draw upon Islamic Ismaili thinks al-Sijistânî and al-Dīn Ṭūsī to short-circuit the aporia between Spinoza and Thomas by way of a very different form of apophatic thought, from the radical transcendence of the One beyond Being and Alterity, at work there. I will end by disempowering this transcendence through Laruelle's conception of non-thetic transcendence in order to finally produce a thought nature (a nature that is thought and productive of thought at the same time). This recasting of each thinker takes them as simple materials where the conception of nature at work can be shown to be "at work" and thus not a reified nature, but rather a manifestation of nature as a thought at work in a wider ecology. This allows me to end with a theory of nature that locates from these three readings three different functions of nature within the single name. In this way a theory of nature produced by this unified theory of philosophical theology and ecology allows nature to take leave from the Nature of naturalism.

An Ecological Reading of the History of Thought

All that was covered in the preceding chapter bears on a new practice of reading the histories of philosophy and theology. Such a practice of

SEPARATING NATURE FROM THE WORLD   169

reading is ecological. Of course there does exist a form of literary criticism that calls itself ecological, but this practice is not the same as the one I am proposing here. For in ecological literary criticism the critic reads certain environmental themes at work within the literature or traces its impact on environmental and ecological thinking. All of which can be useful and interesting and the point of differentiating the two readings is not antagonistic but simply to point out that this practice remains very different from our proposal. For, while there may be something interesting in tracing certain ecological or ecologically friendly themes in Maximus the Confessor or Immanuel Kant (to pick but two random examples), the gambit of our proposal rests on the idea that there is a more radical ecological way of reading the history of thought. Of course, to summarize what I have already argued, the point is that thought is an ecological process. While I'm not proposing a law here, I do take it that this idea is true with regard to something like what is normally referred to as an "empirical claim." That is, though I'm not doing this in any great detail here, thought could be traced in terms of ecosystem functioning at the empirical level. Clearly human beings are ecologically active, constrained by ecological limits, pushing both against those limits and running past them at times, determining the options for how their lives are affected and affect the wider ecosystem and biosphere. To make this claim is not to say that thought is somehow more important than anything else; if an ecologist were to study this idea she may find that it has a weak niche or a weak place in the ecosystem. This would not, however, change the claim, for plenty of species are able to be placed within their nested system without the scientist being required to make a value judgment on that species' place in the ecosystem. It is worth noting that in Chicago there is a research group of ecologists and social scientists well funded by the National Science Foundation who are undertaking the first project of its kind to trace "coupled natural and human systems" in the city. The goal of the project, entitled "Coupled Natural Human Systems in the Chicago Wilderness: Evaluating the Biodiversity and Social Outcomes of Different Models of Restoration Planning," is in part to look at the way different social and economic groups in Chicago have different relationships with their ecosystem.

But I am not just appealing to a scientific research project to shore up my claims by giving it some empirical validity. This would assume already a kind of philosophy *of* science, one where philosophy claims to come along and explain the science we have separated from the transcendental conditions for thought while also calling on the empirical to shore up our claims about those transcendental conditions. The claim of an ecology (of) thought is a claim about the radically immanent character of ecology and thought: ecology as a science and thought as a generic practice that

170   A NON-PHILOSOPHICAL THEORY OF NATURE

takes place in a variety of ways. Ecology as a way of thinking that takes the
posture of thinking from the Real or from a perspective prior and disinterested in the transcendence of Being and Alterity. So what do these claims
do? What does this theory add? They model thought ecologically and this
then makes certain illusions of the Principles of Sufficient Philosophy
and Theology more difficult to hold as well as those illusions of what we
could term the Principle of Sufficient Ecology, for it shows that a thought
that has finished thinking ecologically would be impossible. Thought is
understood not outside ecological practices, but as part and parcel of them.
There is a drive to deal with the ecological crisis by way of something
called ecological awareness, but this seems all too often to remain unaware
of its actual ecological complexity *as a drive toward ecological awareness.*

   Thus, the claim at work here is not something I aim to prove as such.
So, while I suspect that one could do the analysis of thought and ideas in
the way I'm suggesting, I'm not trying to undertake such a purely scientific inquiry. Rather my unified theory of philosophical theology and ecology is directed in this instance toward the philosophical and theological
ideas of nature, to bring them back from the circularity of their faith and
self-sufficiency and show that they are but simple materials. Not "mere"
materials, for as material they are still powerful, but that as materials they
can be used to construct a theory of nature that is both philosophical/theo-
logical and ecological. A theory that does not aim to master nature, but to
show how nature can function in thought in different ways.

   To see how a certain population (of) thought or exchange of energy can
happen consider the common practice in both philosophy and theology of
producing "readings" of other thinkers. Many scholars make their careers
on offering some kind of reading of a historical thinker that allows the
rest of us to read that figure differently than we were able to before theirs.
What happens there ecologically? As already stated, we will not aim to provide the usual history of philosophy, because we can locate the common
practice in philosophy and theology of producing readings as an ecological process. For the reading is the extension of a kind of ecosystem (of)
thought to another locality. The particular shape of that ecosystem will be
different, but its general form will be identifiable. So, for example, the wars
that rage between various forms of Thomisms is not of particular interest
here. While some readings will obviously be wrong in some uninteresting
way, say based on some error of translation or something in the same register, the dominant forms are still populated by populations (of) thought
that find their original organization in the thought of Thomas himself.
At a simple level, the different ecological forms of Thomism exist because
they work, because they can create some niche. In what follows here I will
call this creation of a niche "piety." Piety should here be understood as an

SEPARATING NATURE FROM THE WORLD   171

ecological conception, but only since there is no absolute division between
the realm of the natural world and the realm of religion. Both are in ecological process in a way that changes the fundamental meaning of both
terms. Consider then one of the aspects of Aquinas's thought that has been
recovered by contemporary theologians in all denominations: the concept of
the *analogia entis*. Contemporary Christian theologians tell us magnificent
things about the *analogia entis* such that one could easily come to think it is
the key to Aquinas's whole way of thinking. But such an understanding of
the *analogia entis* removes it from the processes of thought that produced
it. It presents the *analogia entis* not as technology or a method of intuition
that may lead to knowledge and experience, but as a tool of the intellect to
be wielded in the demolition of other processes of thought or the intellectual construction of social and political projects.[1] Such a use of the *analogia entis* has been common since Pope Leo XII's encyclical "Aeterni Patris"
made Aquinas's theological system normative for all theological training.
Such a use of Aquinas turns his concepts into truncheons that the police,
as theologian Mark D. Jordan terms them, then used against those who
challenge their authority.[2] These "police" are more than simply academic
bullies parading the halls of theology and religious studies departments,
for they are also the court intellectuals for a kind of soft (though deadly)
authoritarianism on the rise in Europe and North America. And of course
it cannot be denied that the thought of Aquinas has been used to support a
great number of repressive, authoritarian regimes throughout the Catholic
world, by reinforcing reactionary *doxa* and instilling these regimes with a
sense of divine mission.

Yet Jordan has sought to uncover a relatively "unknown Aquinas"
that challenged the dominant forms of authority.[3] Jordan unearths this
unknown Aquinas through a clear method of situating, in increasing,
intricacy, Aquinas's historical position. Yet, what this really reveals is
Aquinas's piety. Piety is here understood in the sense given it by Philip
Goodchild: "We shall call 'piety' any determinate practice of directing
attention."[4] Piety is the way in which the thinker disciplines their attention, rather than allowing it to be captured. Piety is then an act of liberty
on the part of the thinker that shapes the experiences and networks of
interdependence that are located at the genesis of the thought. For Aquinas
the experiences and networks of interdependence are relatively well-known
by his readers.[5] According to the custom of his time Aquinas was destined
for a life of service to the Catholic Church, as he was the youngest child of
his family. His relatively comfortable family hoped that he would become
a successful abbot of the local monastery. Aquinas, to the chagrin and
sometimes physical protest of his family, pursued a far more humble life

172   A NON-PHILOSOPHICAL THEORY OF NATURE

as a mendicant friar in the Dominican Order of Preachers.[6] Here both his impressive intelligence and desire to serve as Christ had served were given equal opportunity for satisfaction. Aquinas appears to have been very concerned with the concrete realities facing human beings in a time of intense strife and peril as Torrell remarks on Aquinas's sermons: "For an intellectual, Thomas's preaching appears astonishingly concrete, supported by daily experience, concerned with social and economic justice."[7]

The concreteness of Aquinas's thought was surpassed by his intense aptitude for abstract thought. Under the guidance of Albertus Magnus he made an intense study of the Christian scriptures as well as the main theological manuals of the time, but he was also exposed to the work of Islamic theology and philosophy and the Greek philosophy of Plato and Aristotle. This aptitude and appetite for learning led to positions, both at relatively young ages, as a bachelor and later as a master at the University of Paris. There he entered into bitter political debates with the secular members of the university over the right of the mendicants to teach. At the same time his tolerance and qualified admiration for the work of Aristotle often led to conflicts with other members of the Faculty of Theology at the University of Paris. Thus at any one time Aquinas's attention was given to abstract metaphysical debates about God and the World, the concrete defense of the mendicants' religious way of life, and the political realities of a time in great conflict, both inside the church and between the church and varied kingdoms of European Christendom.[8]

This short sketch serves to give an indication of the experiences and networks of interdependence that conditioned the generation of Aquinas's thought. His act of liberty in the midst of these constraining conditions was to focus his attention on speaking truthfully about God and creatures. This means speaking about God in such a way that God is not made into a mere object of human thought, but to speak of God in such a way that his transcendence is respected while also faithfully speaking of God in the indirect and negative ways available to all human persons as natural and rational creatures. It also means speaking about creatures and raising them to the dignity that their existence as creatures made in the image of God speaks to, as well as their possibility of becoming divine as Sons of God. The crystal of piety through which Aquinas was able to unconsciously direct his attention was his doctrine of analogy, specifically the transcendental doctrine of the *analogia entis*. In Aquinas's writings the *analogia entis* appears as a creative line of escape out of certain problems that arise from the inner contradiction of Latin theology's positivist will to truth and its demand for orthodoxy. Which is to say that while the natural philosophy of Aristotle was to be pursued for the truth that rightly belongs to God, as a pagan philosophy it had to be completed by being rethought in Christian dogma.

SEPARATING NATURE FROM THE WORLD   173

Aquinas was able to unravel this contradiction by recourse to the *analogia entis* for the *analogia entis* secures for him the diversity and multiplicity of beings in the natural world without forfeiting, at the theological level, the perfection of God's *ipsum esse*.[9] The *analogia entis* allows Aquinas to secure the radical otherness of God, or God's radical perfection above creatures, without thereby making God unknowable. As Bernard Montagnes writes, "Too close, God ceases to be transcendent; too far, He vanishes into an inaccessible transcendence. In one case He is no longer God; in the other He is no longer real."[10] Thus the *analogia entis* is a process of thought that arises and is able to be productive within a certain ecosystem, but as an ecological process it also by definition has limits.

Or, to take another example, consider the philosophy of Spinoza and the recent revival of his thought in Continental philosophy. The focus of the French and Italian philosophers has been on the *conatus*. Their reading of Spinoza has hinged on proliferating this population (of) thought in their own work, expanding its niche arguably past the limits we find it has in Spinoza's own. For Spinoza, the *conatus* was a population (of) thought that operated but was checked by its coming up against the dominant question of form that determined philosophical thinking at that time. So, for Spinoza, the *conatus* may well have been predominately a population (of) thought in relation to the population (of) thought of the human being (or Spinoza's conception of the human being in more standard terminology), whereas the twentieth-century neo-Spinozists extended *conatus* to *all* things in an intensive way. Spinoza was, of course, excommunicated from the Amsterdam Jewish community at the age of 24 in 1656 and his *Theological-Political Treatise* was banned in the Dutch Republic shortly after its publication due to the pressure exerted on the De Witt government by the Reformed population.[11] This was due largely to the perceived impiety of the book and, indeed, it was (and continues to be) hostile to the dominant pieties that reigned in Europe at the time. These pieties allowed people to war against one another in the name of religion, to persecute men of learning who doggedly pursued the truth of reality without the proper deference to the perceived true cosmology of the Christian scriptures, and ultimately this piety led to the *Ultami barbarorum* that Spinoza witnessed and protested against when an Orangist mob killed and mutilated the recently deposed De Witt brothers.

But, as the standard tales about Spinoza tell us, does this not mean that Spinoza himself actually lacked piety? Following again the philosophical conception of piety that Goodchild gives this word, we can actually see that the perceived impiety of Spinoza was itself an accident of a more fundamental piety. The intentional direction of attention that allowed Spinoza to escape capture by the myriad experiences and networks of interdependence

174   A NON-PHILOSOPHICAL THEORY OF NATURE

within which his life and thought were productive. I will begin by following Goodchild's own analysis of Spinoza's piety. Goodchild begins that analysis by quoting from the *Theological-Political Treatise* where Spinoza locates "the following manifestation of Hebrew piety" writing:

> And here at the outset we must note that the Jews never specify intermediate or particular causes and take no notice of them, but owing to religion and piety, or (in the common phrase) "for devotion's sake," refer everything back to God. For example, if they have made some money by a business transaction, they say that God has stirred their heart; and if they think of something, they say that God has said it to them.[12]

Goodchild connects this location of a Hebrew piety by Spinoza to a Calvinist piety by way of resemblance: "By taking this Calvinist piety to its logical extreme, attributing all that happens to God, Spinoza is able to identify God with nature. [...] Spinoza's method of immanent critique is clear: he began from the ultimate principle, the Word or mind of God, and attributed to it all the properties required by piety, including unity, universality and infinite power."[13] In a similar way to that which Aquinas developed analogy Spinoza developed the thinking of *Deus sive Natura*. Spinoza accepted the dominant pieties of his age by working with them as material to direct attention. They differ in that Spinoza radically took that dominant piety to its conclusion. Or, stated differently, Spinoza's practice of piety is precisely different from that of Aquinas's in that Spinoza's practice of piety effectuates the beginning of the critique of piety by piety itself. Goodchild characterizes it this way, "Benedict (formerly Baruch) de Spinoza, writing at the cusp of modernity, pioneered a method of immanent critique through a cynical equivocation: *deus sive natura*."[14]

Now, as with the debates in Thomism, I am not interested in whether or not this was Spinoza's intention or even if it is the "best" or "most creative" reading of Spinoza. On the basis of the second axiom of non-theology the Real is foreclosed to tradition and authority and thus Spinoza is subject to the Real rather than his philosophy uniquely disclosing it. His philosophy can be revealing of occasions and even help us to work out axioms concerning how to think from the Real, but it is material rather than divine decree. The same is true for the other thinkers engaged with here, including Laruelle whose work nevertheless remains of the utmost importance because of his revealing this structure. Laruelle's work is prized as a particularly useful method and revealing of such structures, rather than treating Laruelle the man as an oracle.

Of course any expert in Aquinas, Spinoza, Ismaili thought, or any of the other thinkers engaged with in this concluding part may challenge

SEPARATING NATURE FROM THE WORLD   175

our use of these figures, but that expert cannot challenge that use on the basis of a reading. I'm treating each figure minimally as a generic name for particular ecosystems (of) thought. Obviously this does not absolve me of scholarly rigor, which I hope the citational apparatus will witness to, but it does free this work as non-philosophy to experiment with their thoughts for a project that is more concerned with a disempowering of the specular culture of reading in the service of the creation of a theory of nature. Not a reading of nature as Aquinas or Spinoza or al-Sijistânî understood nature, but a theory of nature derived from philosophies and theologies as immanent ecological material.

The Dominance of the World

To begin, following our path of a non-philosophical immanental ecology I will consider a particular ecophilosophy that has become a dominant species within the wider ecosystem (of) thought that is Continental environmental philosophy and environmental theology. Within both these fields there has been a focus on the holistic understanding of nature, the idea that there is a single all-encompassing whole, a meaningful whole, of the ecological being-with of the planet. This idea could be rooted in ecological science, as there is a continuum of ecosystems that includes, as discussed in part III, the biosphere. In simple terms, the biosphere is the name for the world ecosystem that contains all living animals and plants and the elements necessary for the possibility of life.[15] The idea, though, of a certain being-with of the whole was popularized by various new religious movements (commonly referred to as the New Age movement) and some theologians within traditional religious groups who, with particular enthusiasm, took to James Lovelock's "Gaia hypothesis." Subsequently scientific ecologists felt at first that there was something amiss in Lovelock's hypothesis, but even though they eventually came to accept aspects of the theory they still felt that the term "Gaia" was corrupted and so they opted for the term "geophysiology." Lovelock's idea was essentially an updated version of the early ecological theories of Clements, also discussed in part III, where the earth is understood as a superorganism.[16] Lovelock's hypothesis is just that, a hypothesis, and as such it is ultimately a heuristic concept that has allowed for interesting advances in ecology. Yet, these advances in the technical literature are rarely, if ever, of interest to the environmental philosopher or theologian, bracketing for the moment any kind of judgment on that lack of interest. Instead, the focus has been on the "holistic" element of Lovelock's theory. This is voiced most clearly by the environmental

176 A NON-PHILOSOPHICAL THEORY OF NATURE

theologian Anne Primavesi in her direct and sustained engagement with the Gaia hypothesis in her book *Sacred Gaia: Holistic Theology and Earth System Science* (2000). There she discusses how her work "gives priority to ecology" in the language that she deploys in her theology, writing that "if theology is to have a positive input into the important environmental debates of the day then theologians have to use or at least to familiarize themselves with scientific environmental language and with its implicit understanding that the ecosystems to which we all belong interconnect within a greater whole."[17]

What is it about this particular idea, this population (of) thought, whose origin appears at first to be from a popular idea about what people think ecology thinks, that allows it to thrive within the ecosystem (of) thought that is environmental thinking (encompassing both philosophy and theology)? The working hypothesis that guides this chapter is that there is a metaphysical commitment that is common to both Christian theology and European philosophy. Genealogists would trace that commitment, obsessing over its genetic element, attempting to tease out the pure source of the idea, but the immanental ecology I have already deployed means that this genetic element is of secondary interest. The fact is that these two ecosystems (of) thought share much in common, including many of the same populations (of) thought that produce different niches within the particular ecosystem. The particular population (of) thought that dominates environmental thought is that of the World. Primavesi locates that similarity between theology and science defining the former as "what we say we know about the relationship between the world and God" and the latter as "what we say we know about the nature of the world."[18] This illustrates what I have already said and will explain in greater detail in this chapter, which is that there is a fundamental relationship, in different forms (dialectical, analogical, univocal, etc.), between God and Nature. What the World provides philosophy is an abstract field where God and Nature become things that are subsumed within a transcendent form of philosophical and/or theological thinking. The philosopher is always above the World as transcendental ego and the theologian is always in the World, but not of it. The ambiguous relationship of philosophy and theology to the World is perhaps best captured by Marx in his famous eleventh thesis on Feuerbach: "The philosophers have only *interpreted the world*, in various ways; the point, however, is to *change* it."[19] The World is this strange invariant and dominant population (of) thought that takes on different functions, is implicated in different relations of energy exchange, within the various ecosystems (of) thought called philosophy, theology, and even science.

Heidegger himself located this primacy of World in a number of his investigations into the nature of philosophy and so there is a kind of worldly

SEPARATING NATURE FROM THE WORLD   177

environmentalism to his work. Because I am claiming that this metaphysical dominance is invariant within philosophy I turn to Badiou, whose opposition to Heidegger allows a short comparative study that shows the varying ways that this invariant dominance of the World expresses itself in considerations of nature.

"Poem or Matheme?": The Worldly Amphibology

Heidegger privileges poetry as the *hors*-philosophical discourse that lets Being be, that allows Being to "unveil" itself to man in his dwelling. He writes, "The most difficult learning is coming to know actually and to the very foundations what we already know. Such learning, with which we are here solely concerned, demands dwelling continually on what appears to be nearest to us, for instance, on the question of what a thing is" and "[p]oetry is what really lets us dwell. But through what do we attain to a dwelling place? Through building. Poetic creation, which lets us dwell, is a kind of building."[20] Heidegger explicitly rejects any kind of thinking of Being in mathematics and by extension modern science generally, claiming that "[t]he mathematical is [...] a project of thingness which, as it were, skips over the things" and "modern science is *mathematical*."[21] This conception of science, according to Heidegger, infected philosophy with Descartes and led to a thinking of the human subject, the clearing of Being as Dasein, along mathematical, rather than "purely ontological" grounds.[22]

In Heidegger's 1929–30 lecture course, published in English translation under the title *The Fundamental Concepts of Metaphysics: World, Finitude, Solitude*, he claims that "homelessness" is "the very determination of philosophy!"[23] This idea is derived from the poet Novalis, the authority of which Heidegger suggests may be suspect from a philosophical position. He goes on to write, "Yet without provoking an argument over the authority and significance of this witness, let us merely recall that art—which includes poetry too—is the sister of philosophy and that all science is perhaps only a servant with respect to philosophy."[24] Thus the homelessness that the poet speaks of is closer to the homelessness of the philosopher, rather than the abstraction from the home of the mathematician. Homelessness is what calls the philosopher forth to speak about the World. What is this World? It is the philosophical urge to "be everywhere at home" when one is not everywhere at home. This urge to be everywhere, Heidegger says, is really a desire to be *within the whole* called forth as a whole: "This 'as a whole' is the world."[25] But, more importantly, there is a population (of) thought at

178   A NON-PHILOSOPHICAL THEORY OF NATURE

work here that even determines this "as a whole," and that is the population of "being." For, "[t]his is where we are driven in our homesickness: to *being* as a whole."[26]

It is in the course of this lecture series that Heidegger touches on a claim he makes in various other writings: that the Greek conception of *physis* does not intend the same meaning as the Latin *natura*.[27] Rather, he claims that for the Greeks the word originally intended sense of *physis* was the "self-forming prevailing of beings as a whole."[28] If the World is "as a whole" and *physis*, which we translate as nature, is "self-forming prevailing of *beings* as a whole," then nature is subsumed into the notion of World. In order to develop this notion of World Heidegger famously drew on the German biologist Jacob von Uexküll, an early protoecological thinker, and his theory of different *Umwelts* for different animals. This was an early conception of the ecological concept of niche that has been greatly surpassed with the discovery of different relations and intensities that form niches, yet the earlier conception of *Umwelt* has captured many philosophers' attention.[29] But of course the metaphysical concept of World does not suddenly appear in the midst of this lecture course, but is of fundamental importance in his 1927 *Being and Time* as well. There, Heidegger shows himself to be a worldly philosopher in the construction of a philosophy ultimately concerned with the worldliness of thought. This is a very different view of his philosophy than he himself put forward, for in his view his philosophy, like all true philosophy, was dealing with the question of ontology, of Being qua Being or the Beingness of beings and so on. In order to get to the discussion of Being, however, he has to pass through the World. Dasein, a first name for human being, has to disclose itself in a World. Heidegger, in his attempt to overturn Platonism, can not simply think Being in the abstract, or as an Ideal subtracted from its actuality, and so his philosophy must quickly come to terms with the World, if not with the state of the World as horizon of Dasein, as that into which Dasein is thrown. Incidentally, it is in this way that Heidegger's thought, as is often noticed, gives itself over to certain environmental concerns. Dasein, the being of human being, does not exist separated from the wider whole of beings, indeed Dasein is said to be in-the-World. Dasein is split between a comportment toward its own Being and "which in each case I myself am."[30] But what is important is not Dasein, but what grounds Dasein and allows it to take on a "definite character" in both modes. Heidegger himself says, "[T]hey must be seen and understood *a priori* as grounded upon that state of Being which we have called '*Being-in-the-world.*'"[31]

What this "being-in" consists of, according to Heidegger, is not a determination of Dasein, but the being-in is a state of Being of Dasein. Thus Dasein is not "side-by-side" with the World, for Dasein is not a thing that

SEPARATING NATURE FROM THE WORLD 179

could be in another thing, but the World is an appearing of Daseins as a whole.[32] Thus, a nonhuman nature is completely wiped out. A sensitive Heideggerian may insist that we say a non-Dasein nature is completely wiped out, but it is clear from the 1929–30 lecture course that Dasein is thought by Heidegger to be the human, hence his faulting von Uexküll for never being "to think thorough the concept of world itself."[33] As well as his famous separation of the human, the animal, and the stone according to their relationship with the world, which he used to uncover the Being of World. Here the human is world-forming, the animal is poor in world, and the stone is worldless.[34] But how is it that the human is world-forming? By tearing away *physis*, through the destruction of the "as a whole," and dragging all of it into the light of truth: "The Greek concept of truth presented here manifests to us an intimate connection between the prevailing of being, their concealment, and man. Man as such, insofar as he exists, in the [*logos*] tears [*physis*], which strives to conceal itself, from concealment and thus brings beings to their truth."[35] Of course the sense of *logos* here is not the same as found in Christian scholasticism, which connects it to a notion of reason that Heidegger claims is enthroned in modern science, but to a simple making manifest or revealing. There is thus a strong relationship between truth [*aletheia*] and *logos* for Heidegger.[36]

But what is most truthful, most revealing of Being, is the world-forming character of poetry or *poiesis*, which Heidegger tells us means "making."[37] This is the thrust of Heidegger's reflection on Hölderlin's enigmatic saying "…poetically man dwells…," which captures Heidegger's attention. For there he writes that the poet isn't one who suddenly "turns up" in the midst of dwelling, but "poetry first causes dwelling to be dwelling."[38] But why is it that the World, which has come to be the name for *physis* and thus subsumes what we may call nature into its worldliness, has to be subsumed into the Being of Dasein in order to be revealed? Doesn't this turn nature, as the World, into a thing in the World? I will return to this question in the next section, but first need to complete the description of this homology by showing how it works in mathematically driven ontology as well.

Alain Badiou is perhaps the most anti-Heideggerian philosopher that nonetheless exists within the same larger ecosystem (of) thought—philosophical ontology. Both philosophers concern themselves primarily with the question of ontology, seeing the question of Being as the primary or underlying question for thought. Both turn to practices outside of philosophy in order to answer that question, but that is precisely where the difference between them lies. Heidegger's ontology is explicitly poetic and Badiou's ontology in contrast is explicitly mathematical, "mathematics *is* ontology—the science of being qua being."[39] Badiou, like Heidegger, is correlatively concerned with the identity of philosophy, but unlike

180   A NON-PHILOSOPHICAL THEORY OF NATURE

Heidegger he does not suture philosophy to ontology alone, which is to say to mathematics alone. For Heidegger philosophy is ambiguous in its essence, seeming to be a proclamation of a worldview or an afterthought of science, but also something absolute and ultimate for Dasein, for letting "Da-sein become what it can be."[40] The identity of philosophy remains, unlike that of science, a problem for Heidegger throughout his works, taking different forms in different periods, sometimes even dropping the name "philosophy" and replacing it with the name "thought," while at other times philosophy and metaphysics are synonymous. In short, the problem remained for Heidegger and as a problem he dwelled with it. For Badiou, however, philosophy is not an autonomous act, in-itself philosophy does not exist, but is dependent upon events that happen in four different realms, which he calls the conditions of philosophy: matheme, poetry/art, political invention, and love.[41] These conditions, which are themselves named truth-procedures, produce the truths that philosophy thinks contingently in relation to that procedure: "Philosophy pronounces, not the truth, but the *conjuncture*, which is to say the thinkable conjuncture, of truths."[42] Thus, though Badiou himself is a philosopher concerned primarily with ontology (though there is a tension between the primacy of politics or mathematics in his work) and thus with a mathematical philosophy, he does not limit philosophy to matheme, which is to say that he does not limit philosophy to ontology.[43] Indeed, what a true philosopher must do, says Badiou, is "propose a conceptual framework in which the contemporary compossibility of these conditions can be grasped."[44]

Therefore, when Badiou poses the question "Nature: Poem or Matheme?" in *Being and Event* (2005) he is actually locating, within the perspective of his philosophy, a *false problem* that plagued Heidegger.[45] Given Badiou's pitching his ontology against that of Heidegger's one might expect that he would simply reverse Heidegger's judgment, perhaps arguing that the poem conceals Being while mathematics unveils it. However, Badiou instead proposes "not an overturning but *another* disposition of these two orientations."[46] That an "authentic originary thought" occurs in the poem is granted by Badiou, but that originary thought is not ontological as such, it is instead a consideration of presence or appearing, which is distinct, for Badiou, from Being qua Being. Thus we may say, following Badiou, that Heidegger confused the philosophy of nature in suturing that philosophy to the poem. I will leave aside the validity of Badiou's contention that Heidegger is wrong concerning the role of *physis* in Greek philosophy, because I am unable to judge it and because it is unimportant within this immanental approach.[47] What is important within our immanental approach is to locate the structure of Badiou's thinking of nature and from that perspective his discussion of the Greeks is a symptom of

SEPARATING NATURE FROM THE WORLD   181

that structure. Badiou holds that the true significance of the Greek birth of philosophy is not poetry, for poetry is more universal and found in prephilosophical societies (China, India, and Egypt are named), but rather what constitutes the Greek event is the *interruption* of the poem with the matheme. The poem that is interrupted is the same poem that Badiou "willingly admits" is the site of the letting-be of appearing a letting-be that is sutured to the "theme of nature." Thus, by implication, the Greek event of ontology interrupts nature as the site of appearance, untying the "thought of Being from its poetic enchainment to natural appearing."[48]

This need to find at philosophy's origin a fundamental privileging of Being and the matheme over nature and poetry discloses the hierarchical structure, which is to say worldly structure, of Badiou's philosophy. Now, it would be misguided, especially within an ecological context, to simply reject hierarchy, for the concept is used though not in the absolute form found in philosophy and theology. But it remains important to note that it is that hierarchical structure that mediates the thinking of the particular "concept of nature," as Badiou calls it.[49] We can extrapolate from here to a kind of formal explanation of his system of philosophy: philosophy is conditioned by truth events that arise from one of the four domains, but philosophy must show how the truth-event is compossible within the other truth-procedures (thus the domain has a fundamentally double structure, both as a world where a truth occurs and as a discourse that pronounces that truth or makes its appearance intelligible); thus the object taken for philosophical analysis (as analysis via the conceptual framework of compossibility) will manifest itself distinctly and fully within this framework and philosophy will pronounce the conjuncture of this object. However, it is clear from the overall form of Badiou's work, especially in his major philosophical books, that, though philosophy is not to be sutured to the mathematical, it is not to take a democratic form between the conditions either. Mathematics, and thus ontology, is thus of ultimate regard and one can almost say, slipping into Laruelle's conceptual vocabulary, that *in the last instance* the object is mathematic.[50]

This is what plays out in relation to the concept of nature in his analysis. Some aspect of nature remains within the truth-procedure of poetry, though Badiou does not elaborate on what that is and his main interest remains ontological asking, "What happens—for that part of it which has not been entrusted to the poem—to the concept of 'nature' in this configuration [...] the framework of mathematical ontology? [...] is there a pertinent concept of nature in the doctrine of the multiple?"[51] The answer is a subtraction from the Heideggerian discussion of nature, where nature as *physis* names that which comes to stand and remain standing in itself, but this ultimately must still be translated into Badiou's own terms

182 A NON-PHILOSOPHICAL THEORY OF NATURE

as a reciprocity between the natural and the normal: "[N]ature is what is rigorously normal in being."[52] Badiou goes on to provide a set-theoretical explication of nature/normal-being as ordinal number and thus nature has been known ontologically through set-theory as "the natural."[53] Ontologically the nature of modern philosophy and science, the Nature of Galileo written in mathematical language, does not exist, because nature as natural-being-in-totality contradicts the axiom that forbids auto-belonging of a set to itself (nature too would have to be a set, as this is where Being qua Being is understood). "Nature has no sayable being. There are only *some* natural beings. [...] The ontological doctrine of natural multiplicities thus results, on the one hand, in the recognition of their universal intrication, and on the other hand, in the inexistence of their Whole."[54]

I propose here a creative rereading of Badiou's metaphilosophy in relation to the question of nature, one that is faithful to the structure of Badiou's thought, but that takes leave with the up to now exegetical character of this section, now casting his philosophy from the immanental ecology. This immanental ecology will disclose the privileged place of wholeness, and thus the amphibology of matheme or poem within worldly thought. Begin first with the general structure of the ecosystem, the exchange of energy between the living, the dead, and the never-living. Such a structure is formally present in Badiou's founding axiom of his later work, which he calls, not uncontroversially, "dialectical materialism." The axiom of dialectical materialism states, "There are only bodies and language; except that there are truths."[55] It would appear then that we have three sites of exchange, but the important difference to note is that while in ecology these three terms are not convertible within the moment of the ecosystem (so the dead is not yet the living and neither are they the never-living), within Badiou's axiom bodies and languages have a somewhat reciprocal relationship yet truths have no such relationship: "These are incorporeal bodies, languages devoid of meaning, unconditioned supplements. Truths exist as exceptions to what there is."[56] We cannot, then, simply overlay the ecosystem concept over the top of the founding axiom, but must tease out the true ecological elements of his thought. First, one notes that within Badiou's ecosystem (of) thought there are seemingly *four* ecosystems, which are his truth-procedures of mathematics, art, politics, and love; however because philosophy is the framework of the compossibilty of these conditions we have to understand these as the dominant populations (of) thought. Within this framework it is clear that mathematics is the dominant species, and that all niches have some relation to this species. The exchange of energy is concentrated here, and if the mathematical species were to disappear the whole of network of relations would fall apart ending in a poor diversity of thought.

SEPARATING NATURE FROM THE WORLD   183

So what then is the never-living condition for the continued health of this dominant species? The only answer available is the World. This has particular relevance for his philosophy of nature, where nature, as a population (of) thought, can no longer function in the niche it has in Heidegger's ecosystem (of) thought as "being as a whole." The niche it comes to take in Badiou's ecosystem (of) thought is as an incomplete Whole of wholeness, or the consistency of a particular multiple.[57] This merely relativizes as appearance, via set-theory, the maximal ontological wholeness of Heidegger's World, but the World is still necessarily the place, the *environment*, in which bodies, language, and truths *appear*.[58] This is clear from his definition of World: "[A] world is the place in which objects appear."[59] Nature, then, can only be understood via World within this ecosystem (of) thought and thus, Badiou, like Heidegger, subsumes the real identity of nature into World in a way that is structurally the same.

Unilateralizing the Fourfold

Let us return now to the question posed earlier: why is it that the World, as the name Heidegger claims is closer to the intention of the Greeks' meaning of *physis*, must be subsumed into the Being of Dasein in order to be revealed? And doesn't this turn nature, as subsumed into the World, into yet another thing in the World? The answer seems to me to be yes, but there is an important corollary to this thingness of nature subsumed into World: there is a reversibility between the World and Dasein's Being, which is in fact the genesis of the thingness of nature at work here. For the World is the appearance of Being and Being is what produces the appearance such that Heidegger's thought continually comes to turn around these two terms throughout his work.[60] This becomes clearest, perhaps ironically, in Heidegger's conception of the fourfold. The potential irony arises from the fact that, of all Heidegger's concepts, the fourfold is the most poetic and thus the biggest distraction for Continentalists in their attempts to have mainstream Anglophone philosophers take seriously the work of Heidegger. For what does it mean to say that "[t]he fouring presences as the worlding of world"?[61] Especially when that meaning is dependent upon the "fouring" of "earth and sky, divinities and mortals"![62]

Yet, while of course the fourfold is indeed strange at first glance, it still works as a way of thinking about the presence of things. It is through this fourfold that one who is thinking is able to presence the thingness of some thing, to speak Heideggerian. The power, in terms of its ability to produce thought, may become clearer if one thinks of the fourfold as a particular population (of) thought whose niche it develops parasitically to two other populations: World (with nature subsumed within) and Being, or "significant appearance" and Being as such.[63] What it produces, then, is a mixture within thought of World and Being, which nature is subsumed within. Thus, we aim to unilateralize the fourfold in order to separate out nature from World. Precisely in order to think nature prior to Being and Alterity, that is, nature as in-One, we need to disempower this particular population by changing its niche, allowing nature as in-One to become separate from "significant appearance" or World.

It is necessary to explain what the fourfold is aside from its poetic form given it by Heidegger. First, we know that the conception of the fourfold flows out of Heidegger's obsession with poetry (what could be called the space and time of his ecosystem (of) thought). It is this poetic aspect of the conception that has likely scared off most of his commentators from dealing with it, since, as mentioned earlier, it is already at odds with attempts to translate Heidegger into more acceptable forms of philosophical thinking. It is unsurprising then that in English I know of only three secondary sources that take the fourfold seriously in their reading of Heidegger's philosophy. Seriously is meant here in the sense that they use the fourfold as a way to understand Heidegger's philosophy as a whole and so it works as a kind of surveying matrix arising out of his philosophy that can help organize a reading of it. These thinkers are Julian Young in an article added to the second edition of the *Cambridge Companion to Heidegger* (2006), Michael Lewis's *Heidegger beyond Deconstruction: On Nature* (2007) where the fourfold is used to explain Heidegger's philosophy of nature, and in Graham Harman's *The Quadruple Object* (2011) where the fourfold is subtracted from Heidegger's phenomenology and used in the service of a philosophy of objects. What is interesting is that each author, while taking the fourfold seriously, seemingly presents a very different interpretation. Yet, despite these differences, each author is also compelling in their reading and supports it with an intimidating scholarly apparatus. I am ultimately uninterested in the scholastic nature of Heidegger research necessary for any attempt to construct a judgment on which is the best and most accurate retelling of the fourfold. Indeed, following the non-philosophical method I've been developing in this book, I am able to treat each of them as a generic ecosystem that exhibits fourfold-like properties, simply because each of them *does work*.

Each reading works because they are each pitched at different scales, remembering that the sizes of ecosystems are always *n*-dimensional. Thus for Harman the fourfold maps an object into a duel that takes place at two levels: one of presence and absence and another of veiling and unveiling.[64] This is because a thing, or object in Harman's vocabulary, is both

SEPARATING NATURE FROM THE WORLD 185

within the world as an individual being-as-such as well as breaking or tearing apart the being-as-such by veiling being-as-a-whole.[65] This zooming into objects as such is understood at a different level by Young, where the fourfold maps things onto the nature/culture difference. This is because the axis of divinities and mortals has to do with the realm of culture (in terms of values and finitude) while the axis of sky and earth has to do with nature.[66] And finally, for Lewis, the fourfold has to do with signification of nature and thus it is a level below Young in its combining of nature and culture within the naturally formed act of self-transcending through language.[67] These differing levels are helpful for our purposes insofar as what Harman's use shows is that the thing that appears within the fourfold is a simple object in relation to other objects. This reveals something that is more fully expressed by Lewis, as the fourfold is the presence of the thing as signification within the World.

So what then does it mean to unilateralize this fourfold? Keeping in mind that the fourfold is a production of thought, so the fourfold itself is not the ground of thinking but an organization of the movement of thought, then a unilateralization of the fourfold is simply another way of organizing thought. If for Heidegger the fourfold is about appearance, the unilateralization of the fourfold makes such appearance relative to its real identity where there can be no separation between its static image and its movement. Laruelle's conception of unilateralization at work in non-philosophy serves to direct thought from the One, to direct thought toward the most radical immanent form of what is thought. Thus, in unilateralizing the fourfold I hope to organize thinking in such a way as to free a thought nature from reversibility or the dialectic of Being and Alterity. Nature is not an other to humanity, nor is it the field of Being, but has an identity. Heidegger's focus on measure in "...Poetically Man Dwells..." suggests that the fourfold is concerned with a kind of individuation as taking the measure of a thing. Yet, nature is perverse, it is, as One, immeasurable. Thus any thinking of nature must be immeasurable, without recourse to transcendental circumscription, dialectics, or fourfolds.

While I have drawn on the works of the various commentators already discussed in order to form my understanding of the fourfold, here I present a different understanding of its various combinations. First, the two axes can be interpreted ecologically. The axis of earth and sky can be thought of as the biosphere upon which the second axis of mortals and divinities, or death and life, is dependent yet inextricably linked. That is, the earth and sky is a reduced duality that speaks to the fundamental level of the space of the biosphere (earth) and the rhythm of the relationship of that space with the energy provided by the sun alongside of other relatively transcendent elements to the earth like climate (sky).[68] This is especially clear insofar as

186   A Non-Philosophical Theory of Nature

the earth is what is seemingly absent in its closeness, while the sky is where the unveiling of transcendence happens. Climate becomes clear to us not as a whole, but in our experience of weather as manifest in the unexpected storm on the horizon or the haze of an especially hot Chicago summer day. The second axis is the life and death that are sustained by this first axis, yet also constitute its actuality. For the mortals are those who "are the human beings. They are called mortals because they can die. To die means to be capable of death *as* death."[69] The divinities are those who provide the measure of that unthinkable moment, a moment that mortals witness, but do not experience in that witness. That is, the divinities are the measure of this moment through their absence, which Heidegger calls their default, which presences a hidden fullness.[70] This hidden fullness is only experienced because of mortal finitude, for Heidegger, because of the fact that an individual human being is not the fullness of being.

This then creates an interesting new set of binaries that are more abstract and that are lacking in the three commentators we've discussed: that of earth and divinities and sky and mortals. For there is a deep connection between the veiled character of the earth and the default absence of the gods at work in the fourfold, just as there is a deep connection between the presence of mortals and the unveiling of the sky. For, as we stated earlier, the fourfold is ultimately concerned with individuation by way of measure. Consider Heidegger's words when he writes, "What is the measure for human measuring? God? No. The sky? No. The manifestness of the sky? No. The measure consists in the way in which the god who remains unknown, is revealed *as* such by the sky."[71] Thus, contra Young's suggestion that earth and sky form the axis of nature and mortals and divinities form the axis of culture, it is actually that the quarter (rather than axis) formed by earth and divinities is where nature is given as absent and veiled and the quarter produced by sky and mortals as present and unveiled is where culture is: culture reveals nature *as* such for Heidegger.

Following this quarterial vision we can thus see the fourfold not as fourfold but another duality of identity expressed as (earth and divinities) and (sky and mortals).[72] If, following Laruelle, an identity is always what is One in-the-last-instance, then this duality must be thought unilaterally. But since this duality is in actuality a set of two dualities, the act of unilateralization will require three operations. First unilateralize the duality of the two sets ([earth/divinities] and [sky/mortals]) and then deal with the real identities at work in these two sets. In-the-last-instance the set of sky and mortals is determined by the set of earth and divinities. That is, culture is determined in the last instance by nature and culture has no real effect on nature as such. For, even if culture plays its part in the destruction of the biosphere, it does so naturally. Nature remains perverse in the face

SEPARATING NATURE FROM THE WORLD   187

of culture and quite explicitly nature's perversion is found in the very existence of culture as it is at work naturally in the biosphere. Yet the thinking
of nature in this dyad is caught. Always a thought of two terms in relation: earth and divinities in Heidegger's terminology. These two terms can
stand in for a number of different terms in the history of thought and the
relationship between the two terms varies depending on the system at play.
The idea of "Nature," with the capital N intended, is often aligned with
Heidegger's earth. This is Nature of philosophies of immanence as well as
naturalism. It is often thought in relation to God, aligned with Heidegger's
divinities in this case, and is the God of theological thinking as well as any
form of philosophical transcendence that claims to be outside of the conditions of what is (so of course simple change is not transcendent in this way).
It is by looking at these terms, Nature and God as understood to be generic
terms related to Heidegger's earth and divinities, that we will unilateralize
this dyad or think nature as in-One. We will begin thinking through the
relations of these two terms ostensibly by looking at the way they function in St. Thomas Aquinas, before turning to Benedict de Spinoza and,
breaking with what might seem like a historical survey, turning to Ismaili
thinkers Abu Ya'qûb al-Sijistânî and somewhat on Naṣīr al-Dīn Ṭūsī. But
by engaging with these historical thinkers I am not aiming to provide a
definitive reading. While I have made traditional studies of these thinkers in preparing for this chapter, I do not present these as studies as such.
Instead, those studies have allowed me to treat these thinkers minimally,
as occasions that provide simple material with which to work.

Thus the theory I put forth here is neither Thomist, Spinozist, nor
Ismaili. What we have done here is change the fourfold from where the
"World worlds" and turned it into an n-dimensional ecosystem. We have
shown that the World, an image of representation, is separate from the
ecosystem, which is always an identity of flowing energy amid identities
that are radically immanent in the ecosystem. For now the claim is that the
set of earth and divinities is unilaterally related to the axis of earth and sky
(biosphere), while recognizing that an element of this set is found in the
axis. By unilateralizing the fourfold in this way we have actually treated
it as an ecosystem such that the set of earth and divinities operates as the
living, the dead (earth), and the never-living (divinities) aspect of the ecosystem. This thesis will appear as if it were standard reductive naturalism,
and this is why there is a need to work out the real identities of earth and
divinities as they are found at work in other philosophical and theological
attempts to think nature. In a certain way what we're practicing here is an
ideational "forcing" or *forçage* in the French. This is usually ascribed to
a technical aspect in set theory related to showing how all terms can be
generic without having to do the infinite equations required to actually

derive such a result. But the term also refers to a more obviously ecological practice of forcing a particular plant to grow when it wouldn't be able to. This may involve planting and growing the roots late in the season, as in the summer, and then taking the roots out in the fall to be forced or grown indoors out of season. Or, more commonly, greenhouses are another example of forcing. In each case the plant is removed from its wider ecosystem, such that the soil it grows in is not the same soil that grows in the front garden and so it is not using the nutrients in the same way it would. In the next chapter we will practice this in relation to the four terms discussed before turning to bringing these newly identified terms together again.

Chapter 12

Materials for a Theory of Nature

Materials, Energies, and Populations (of) Thought

A theory of nature that is thought separately from the World must be both non-philosophical and non-theological. It must be both for the theory and must avoid the trap of a Principle of Sufficient Philosophy that would circumscribe nature as a thing captured and examined by its own thought. It must also avoid an all-too-easy reversibility with God. In this way the hypostasized Nature of naturalism is a form of theological thinking. So, the theory must be non-philosophical in order to think nature as in-One, radically autonomous and foreclosed in-the-last-instance to thought, which nevertheless remains relatively autonomous, and must be non-theological to avoid the Principle of Sufficient Theology where the relatively autonomous thought becomes a mere echo of authority, not only of the standard theologies of Worldly religious authorities, but also the unacknowledged theologies at work in those naturalisms that aim to "mirror Nature." The practice of thinking outside these specular forms of thinking, with their complementary forms of self-sufficiency discussed already in part I, is the unilateralization of this dyad of earth and divinities.

This dyad is the mixed-up thinking of nature, always in some relationship between God and Nature, for theology and philosophy. We can demonstrate this by looking at Aquinas and Spinoza as paradigmatic examples of theological and philosophical processes, respectively. Again, not as different truncheons ready at hand for the police (both theologically, philosophically, but also politically), but in its ecological process as an expression of piety.

Creation as Apophatic Name of Nature

Aquinas's thought, following that of Pseudo-Dionysius, undertakes the theological task of speaking about God as God in an indirect and negative, or apophatic, fashion. Yet the form of such an apophatic thinking is given in Aquinas's understanding of veiling and unveiling, of absence and presence. In other words apophaticism is directly related to nature and God. This conception of the relationship between nature and God is encapsulated in his understanding of the doctrine of creation where the veiled (God) covers over and determines nature (absent or unconscious as lived experience in the everyday) for nature only has positive being insofar as it points toward what will always remain veiled in itself. Creation comes to be the locus for an apophaticism, not of the Divine, but ultimately of nature itself. Such an apophaticism is not theistic in the usual sense, for in mainstream philosophical argument theism tends to portray a God that typically looks a great deal like *a* being or absolute entity that exists independently of the rest of reality. This theistic conception of God tends to think of creation as something added on to thinking about God, and while Aquinas clearly sets up an absolute hierarchy between creatures and the Creator, all knowledge of that absolute hierarchy and thus all knowledge of that Creator are dependent upon nature being creation. Creation is a truly nontheistic theological conception of the ontology of nature insofar as it exists in cognitive tension as rationally understandable through a relational or ecological reason and at the same time understood under the strictures of faith as attested to in the biblical revelation of an absolutely transcendent God. This notion of Aquinas as a nontheistic theologian in the tension between ecological reason and biblical faith follows on from the work of Rudi te Velde who states that, for Aquinas, the concept of God spoken about theologically and philosophically is truly "the concept of the relationship of God and world, conceived as an ordered plurality of diverse beings, each of which receives its being from the divine source of being. For Thomas there is no thinking of God concretely outside this relationship."[1] This relational form of thinking is the proper way to understand clearly Aquinas's metaphysical formulation of the divine reality or action of God as *ipsum esse per se subsistens*, or self-subsistent being.[2] For, as Velde's work allows us to see, such a formulation is not strictly speaking a definition of God but of the relationship of God and world or creation. Aquinas could have simply stated *ipsum esse* (being itself), instead this being, inherent to all things that are, is also the *ipsum esse* of the cause of all beings. Thus to think God is to think, for Aquinas, not abstract being as such but that simple abstract being in relation to the diversity of perfections that are

MATERIALS FOR A THEORY OF NATURE 191

unified in their cause, which is that simple being.³ Thus, at the heart of his doctrine of God is creation.

With regard to the human capacity to understand God, analogy allows one to recognize the epistemological limits in thinking about God as transcendent, but also to locate what can be known through an investigation of God's effects. This doctrine takes its mature formulation in the *Summa Contra Gentiles* where, before moving into the argument for the doctrine of analogy, Aquinas affirms the negative theology of Pseudo-Dionysius at the same time as he affirms that names can be predicated of God.⁴ He does this through a differentiation of meaning and signification. The mode of signification, thinking of God in relation to other things we call good, is imperfect for those good things are never good in themselves. They are related to God as cause or as that by which they are as effect. This means that when we predicate God by the name "good" we have predicated an imperfection in the mode of signification. The term that has the status of creation in the analogy is imperfect, as its being good is dependent on the other term of the analogy—God.⁵ In this way the thinking of any created thing relates us immediately back to God via their imperfection as effect and not cause in itself. To understand how something is good we have to transcend *what they are* and move to *that by which they are*. Then, in order to understand God at all we recognize that the imperfection of the thing that has goodness does not befit God, but the meaning in some eminent way does befit God. With regard to such names that hold in their positive nature imperfection, such as a good thing, Aquinas writes, "such names, therefore, as Dionysius teaches, can be both affirmed and denied of God. They can be affirmed because of the meaning of the name; they can be denied because of the mode of signification."⁶

There are other names that Aquinas says are said of God alone, as they express perfections unavailable to creatures, and belong to him through the mode of supereminence.⁷ This mode of supereminence can only be expressed through negation, as when we say that God is in-finite, un-composed, im-mutable, or un-divided.⁸ These supereminent expressions were not thought by Aquinas to be found in the sensible world and in that sense they cannot be investigated for a further understanding of God's being. We may even go so far as to suggest they are but a matter of fact, but a fact that stands against further positive knowledge and only allowing knowledge as wonder. The desire caused by this wonder may be assuaged by recourse to an investigation of the relation creation has to God, "as when [God] is called the *first cause* or the *highest good*."⁹ This is the indirect way of understanding God, through the divine action in creation. Thus we are given the full apophatic structure of Aquinas's thought in the preamble to his discussion of analogy: "[W]e cannot grasp what God is, but only

192   A NON-PHILOSOPHICAL THEORY OF NATURE

what [God] is not and how other things are related to [God]."[10] This is
then a two-pronged apophaticism—a negative knowledge of what God is
not and an indirectly positive knowledge of God that devastates the idea
that creatures have positive being in themselves.

For analogy to work as a tool of piety it must have both aspects of this
apophaticism. Only in this way can it form thought and direct attention
in a way that satisfies the demands placed upon Aquinas by his situation.
It must give attention to the dependency of humanity, as well as the rest
of creation, on a God who is perfect being itself. At the same time it has
to give attention to the true understanding of God of which humanity,
and perhaps the rest of creation, must have some access other than (non-
existent) pure revelation. As Aquinas understands them, the other grammars or conceptions of how human persons may name and thus think
about God, as qualified by the Christian creeds and Church Fathers, do
not give sufficient attention. Thus univocal predication is not sufficient in
Aquinas's view because it both implicitly denies the radical otherness of
God and attempts to give access to knowledge of God through some third
term. The radical otherness of God is not respected in univocal predication because it attempts to find some commonality between God and
creatures through either a genus, a species, a difference, an accident, or
a property.[11] The third term that allows access to a certain knowledge of
God also removes the difference between God and creation by subjecting
them to a concept that is simpler than the other two. Aquinas affirms
the demands of orthodoxy on this point and writes, "Now, there can be
nothing simpler than God either in reality or in concept."[12] If there were
a concept simpler than God it would mean there was a concept that has a
perfection that, by ontological definition, can only be said to belong, via
supereminence, to God.

Equivocal predication gives insufficient attention to the relationship
of God and creatures, as it allows no understanding of an order of cause
and effect between two terms.[13] For there to be such an understanding a
thing must have some understanding of its own identity, what Aquinas
calls "likeness," in itself, whereas equivocal predication erases this identity through the unity of the name.[14] This identity, or likeness to God
through the relationship of a cause to its effects, is necessary for there to be
any reasoning of God at all. Since creatures are not God and thus cannot
know God as Godself they must proceed in their reasoning from creatures, which share a likeness to God as God's effects, to God, as cause.[15]
In this way equivocal predication of God, as Aquinas understands it, gives
no access to any knowledge of God in much the same way that univocal predication does not. Equivocal predication tells us nothing because
it posits a unity of name that erases the relationship of creatures to God,

MATERIALS FOR A THEORY OF NATURE   193

thereby misdirecting attention away from the dependent being of creatures to pure linguistic-metaphysical speculation. Univocal predication, on the other hand, posits a third term, which contradicts Christian piety as well as ontological definition by erasing the names that name God's perfection and constitute our negative knowing of God.

For analogy to direct attention in the way Aquinas demands it must limit itself to two terms: God and creation.[16] This relational knowing allows for the aforementioned two-pronged apophaticism. First, it recognizes the epistemological strictures as determined by God's perfection that determine that we can only know what God is not. At the same time it locates the access of indirect knowledge to God through creatures. As creatures are not perfect it follows logically that words we ascribe to creatures, such as "good" or "alive," apply to God as perfections upon which creatures are purely dependent. Aquinas says as much when he writes: "Thus, therefore, because we come to a knowledge of God from other things, the reality in the names said of God and other things belongs by priority in God according to [God's] mode of being, but the meaning of the name belongs to God by posterity. And so [God] is said to be named from [God's] effects."[17] Analogy is thus a matter of our understanding or sense of reality. Reality here refers to that which is understood both in terms of the contemporary *doxa* as mundane or the "mere" of everyday life, and as the divine reality of God and God's outworking love in creation. It is thus very tempting to think of Aquinas as a sort of Kantian *avant la lettre* in that analogical knowing appears to be purely epistemological, setting limits around that knowledge, and giving us some access to reality. This reading has been propagated by some analytic readings of Aquinas on the question of being.[18] There exist resources in Aquinas himself to suggest another reading that would align him with what I characterize as ontologically significant apophaticism. But what is meant by ontologically significant apophaticism? Essentially it presents reality as always having a remainder beyond what can be consciously thought or experienced, and says that this remainder is real and not just an epistemic weakness in thinking. In this way it is kataphatic as apophatic, as the apophaticism does not negate reality, or its parts, but speaks to something really beyond this negation, without reifying that beyond as a thing.

The beginning of ontologically significant apophaticism begins with the recognition of epistemological limits to reason already held by apophatic theologians, in a different form, before the Kantian critical philosophy. This limit itself has a limit, for we are able to know some aspects of things with certainty, or else even apophatic knowledge would not be possible. Aquinas locates this in his commentary on Boethius's *De Trinitate* where he distinguishes between "knowing *that* and knowing *what*" writing: "We

194   A NON-PHILOSOPHICAL THEORY OF NATURE

know *what* something is when we understand the essence of it—what makes it what it is—either directly, or [indirectly] by way of something that sufficiently displays what it essentially is."[19] This kind of knowledge is possible, in Aquinas's view, when our senses grasp the thing but not when only the mind can grasp it. This position has the consequence that we cannot know immaterial substances such as God except by analogy from those we sense. As is becoming a familiar trope, and one that is decidedly important for understanding Aquinas's thought, we precede to knowledge of God through knowledge of creation. To be clear Aquinas is claiming that we can know what something is—an apodictic knowledge of its essence—if it exists in the realm of the sensible, but analogy cannot allow us to know the highest within thought apodictically: "so analogy from substances we sense can't lead us to know immaterial ones sufficiently well."[20] It would appear to some that Aquinas is destroying the possibility of any secure theological knowledge. Aquinas himself writes, "During this life then we have no way of knowing what these immaterial substances are, either by natural knowledge or by revelation."[21] However he safeguards a point of secure theological knowledge by arguing that though "we are left not knowing what such immaterial forms are" we do know "*that* they are, whether we know this by natural reasons' arguments, from created effects, or by revelation's use of analogies drawn from what we sense."[22]

This is more than epistemological humility, for when we connect it to the doctrine of analogy it says something significant about the reality of nature itself. This is clear as Aquinas's doctrine of analogy is connected inextricably to his metaphysical account of the divine causality in creation. Velde sums this up succinctly when he states: "Analogy is meant to articulate the commonness of effect [creation] and cause [God]: the effect is *differently the same* as its cause, precisely insofar as it is being."[23] The first instance of this connection between creation (as Nature) and analogy is found in a short text entitled *De principiis naturae* [On the Principles of Nature], which Aquinas wrote at the age of 25.[24] This text was a short exposition on causality that is largely dependent on the Islamic Aristotelian philosophy of Ibn-Rushd.[25] As this is a text on the principles of nature it can only be understood as an ontological text, while at the same time it demonstrates a deep recognition of the epistemological strictures present when a human being attempts to understand not only God, but even nature.[26] Some interesting aspects of an early conception of analogy come out in this text, due in part to it being developed as a largely secular text. So we find that it does not once use the word "God," but ultimately still leads to the theological insight that gives to God the place of cause of everything. So even though the word "God" does not appear in the text, his formulation of causality and end or goal clearly applies to his understanding of the divine action if

MATERIALS FOR A THEORY OF NATURE   195

read in the light of his later works. After running through defining expla-
nations of the three principles of nature (matter, form, and lack of form)
Aquinas turns to writing on "ultimate matter" or "first matter."[27] Ultimate
matter is defined paradoxically by Aquinas in this way: "Only material
subject to form and lack of form but having no particular form or lack of
form in itself can be called *ultimate matter*, because it presupposes no other
material; and another name for it is [*hyle*]."[28]

In Aquinas's account of nature's three principles ultimate matter appears
as a remainder, in that it cannot be thought rationally like the other three,
but can nevertheless be argued for as underling reality. The only reason
that ultimate matter may be thought at all is because of analogy. He writes,
"But since we define and know things by way of their forms, ultimate
matter can't be known or defined as such, but only by an analogy, as that
which relates to all forms and lack of forms as bronze does to statues and
to shapelessness, and so is ultimate."[29] That is, as Aquinas shows us the
principles of nature through analogy with intentional human production like statue making, the mind may easily begin to think that there is
some ultimate kind of potential matter like bronze is for the statue. Some
would suggest that this is an illegitimate anthropomorphism of nature, but
Aquinas is being quite ecological here as human beings are also natural and
thus their actions can be understood to be part of wider natural causation.
On a further ecological note, ultimate matter cannot be known directly as
what it is in itself but only through its relations with other things. Analogy
may then be defined as knowing through relation. Ultimate matter can
only be speculated about via an analogy with material that does have particular form and lack of form and thus Aquinas can make claims about it,
but nothing about it in itself other than what is given by logical necessity.
These claims are ultimately speculative, even little leaps of faith, for we
know individuation or generation through forms and thus, Aquinas leads
us to think, within our own limited ability to know, we have to say that
"[n]othing actually existent then can be called ultimate matter."[30]

But to return to the question of whether this relational knowledge
occurs through some agreement only in one's understanding or in the
wider reality outside that understanding, we note that to know *that* something is "implies some vague knowledge of *what*" it is.[31] Aquinas is using
the notion of analogical knowledge, where knowledge is always connected
to the truth of its being true, to mediate between totalizing knowledge and
equally totalizing ignorance. Furthermore it suggests that knowledge
through analogy implies the truth of being analogically. All knowledge,
in Aquinas's thought, must ultimately follow the vacillation of theological knowing—between knowledge that leads to possession of being by
its being known and ignorance, which leads to vice. In this way all true

196   A NON-PHILOSOPHICAL THEORY OF NATURE

knowledge has an analogy with Aquinas's conception of God's "rustic" knowledge of real singulars.[32] God's knowledge comes not through its being related to real singulars, though there is a relation, but through a proportionality of God's being as cause.[33] God's knowledge cannot come through relation because if God were to know through an absolute relation rather than through a proportion of God's own being, then this would suggest that God was subject to relationality rather than relationality being subject to God. Thus, following the circular logic of analogy, our own knowledge similarly "corresponds" by proportion between what is given in the mind and what is true: "Because for Aquinas truth 'corresponds' not by copying but by a new analogical realization of something in the mind in an inscrutable 'proportion' to how it is in reality."[34]

Following this notion the analogical reasoning that leads to ultimate matter proceeds to the idea that, in reality and not merely in our understanding, nature is not reducible to matter. Setting aside the hylomorphism present in much of Aquinas's writing, which is suspect in light of modern biological and physical sciences that Aquinas could not have known, we can still locate a significant contribution to the explication of an ecological ontology of nature in an apophatic mode. Matter, it would be expected, explains a great deal about nature. Yet, despite our growing understanding of what matter is, we are not able to understand some of the most basic aspects of nature through matter. Is this lack in our understanding due to the epistemological limits of thinking or do those epistemological limits on one aspect of nature, human understanding, speak to a certain *what-ness* about nature itself? Nature is not reducible to matter precisely because such a matter, which would be ultimate, is itself not locatable in actuality as thought and must exceed itself through some form of otherness in order to actually exist. This transcendental claim is repeated at the level of the ecosystem where matter can be located *within* the ecosystem while the ecosystem itself is not locatable simply from the exchange of matter.

It follows that nature understood as reducible to matter would be to think about nature through only one aspect. Ecological reason tells us there must be some relation in order to conceive of nature at all. Nature understood through matter must be conceived in one of two ways, either nature is only material and thus does not make up all of reality or nature is ultimate matter. As ultimate matter has already been shown to not exist in actuality it follows that we cannot conceive of nature in this way. Yet, the first way of understanding nature, as only material, fails to provide a concept of nature that is not mere tautology. If nature is merely material, then why speak of nature at all, and instead why not just speak of material? Rather, the *thatness* of nature as not reducible to matter speaks to the significant, but not yet adequately understood, *whatness* of nature's

MATERIALS FOR A THEORY OF NATURE   197

real, constant exceeding of matter. Nature can thus be given the apophatic name "creation" to express this reality.

As an apophatic name of nature, creation bequeaths a powerful notion to the thinking of nature in philosophy and theology in ontologically significant apophaticism. However, as purely Thomistic such ontologically significant apophaticism has its own limits, which are located in its thinking from analogy. While this section attempts to show the strength of Aquinas's thinking from analogy it must also return to the use of Aquinas by the police (again, following Jordan's terminology) and ask what makes Aquinas's thinking so easily amenable to such reactionary use. The trade-off for giving a certain divine dignity to creation through apophatically thinking it as God's effects is that God's transcendence as limit to violence is undermined, though not destroyed. That is, in non-theological terms, thinking from analogy suffers from its own version of the naturalist fallacy. Analogy must proceed from what is in order to understand the nature of the divine. In doing so it lacks any kind of organon for selection and thus may select, as Aquinas himself did, an analogy of monarchy: "[W]hatever is in accord with nature is best, for nature always operates for the best. But in nature government is always by one. [...] Among the bees is one king bee, and in the whole universe one God is the Maker and Ruler of all."[35] Here Aquinas proceeds not from an understanding of nature to a properly analogical understanding of God but from what human government is to a misunderstanding of what government in nature is and then to a conception of God's governance. Bees, we now know, do not have any one ruler as the queen of a particular hive does not direct the action of that hive; the hive, rather, proceeds in a way altogether unlike human government from monarchy to democracy. Indeed, what is in nature may be best, because it is what is, but what is in nature is varied. Its organization is not reducible to any one organization and thus analogy may not find a secure position from any one part of nature and may not be able to think from the whole of nature in ways that allow it to remain orthodox.

The problem with analogy is then that the entire process of analogical predication is unable to function outside of the purely metaphysical. In terms of a doctrine of God the apophaticism of creation is productive, but it is only productive on the basis of a second-order negation of nature. I have shown that when one selects from what is (nature) to know God, subsequently, that which is (the natural thing) is shown to truly be relative to the *ipsum esse per subsistans*. We first know by way of something natural, but that knowledge is then perfected in the abstract thinking of God thus negating the autonomy of what is natural in the light of its cause. This apophatic thinking of creation ultimately pulls the ground out from under the one thinking, for what the perversity of nature resists is the selection

198   A NON-PHILOSOPHICAL THEORY OF NATURE

of a part of nature. Thus what we ultimately have in Aquinas's seemingly complete system of theology is actually but an ecosystem named creation that is characterized by a minimal and rigid exchange of energies between God and creatures that says nothing positive about nature as such. This analogical ecosystem can only produce a negation of nature and ultimately, in so doing, provides no way of thinking positively about God outside an abstract metaphysics.

Here we see both the theological erasure of nature and its reversibility into a naturalism unmoored from the duration of lived nature (which is of course from the exchange of matter and energy between the living, dead, and never-living aspects of the ecosystem). For analogy is unable to select one aspect of nature with which to speak of God, instead collapsing into a perpetual loop of selection creating a relatively unstable system that is easily disturbed. It is Spinoza who is able to finally think a consistency to God that runs from the metaphysical to the ethical by selecting the variation in nature.

The Chimerical Immanence of God or Nature

Spinoza has been much maligned by the theological establishment, both of his own time and of the contemporary era. So it is no surprise that there is an incredible difference between Aquinas and Spinoza, but it is a difference that goes deeper than simply their different historical situations or even their different positions within faith. The difference lies in the fact that if Aquinas had no organon of selection and thus analogy became a perpetual loop of selection between God and nature, Spinoza instead selected nature in its infinity (which should not be confused with the nature's status as perverse in this work). This is what Goodchild called Spinoza's cynical equivocation of God and nature, where upon asking a metaphysical question he instead gives a material answer. Again the play of veiling and unveiling, presence and absence is vital here. For when someone demands that veiled, God, be unveiled and our attention directed toward that unveiling, Spinoza claims to practice that unveiling by directing attention toward what is absent or unconscious, nature as the lived experience of the everyday. In response to demands to give attention to transcendence, Spinoza directs attention to immanence and claims the two are equivalent. Unveiling remains absent, but neither is God veiled now nor is nature present in a way that would allow an individual to circumscribe it. The veil is seen and it is seen in nature. Yet this cynical equivocation is more powerfully seen not as a rejection of the metaphysical, as Diogenes the Cynic's own gestures could

MATERIALS FOR A THEORY OF NATURE 199

be interpreted, but as the creation of a chimera between God or nature, between a lofty metaphysical question and what matters in the everyday. François Zourabichvili locates the chimera at the heart of Spinoza's thought, existing in a tension between a critique of the chimerical tendencies and a strategy of the chimera as a constitutive practice of that thought. Thus Spinoza both critiques those who peddle snake oil, exploiting the confusion caused by chimerical reasoning, and claims that chimeras and metamorphoses are the ground of thought.[36] Zourabichvili defines the chimera as that "whose nature veils [*enveloppe*] an open contradiction" and that which "by its nature, cannot exist," going on to write that "[t]he chimera is not a thing but, if we can put it this way, a non-thing, a non-nature."[37] In a brilliant use of Zourabichvili's reading, Rocco Gangle argues that the positive strategy of the chimera creates a system-dependent meaning of a common word that reveals the system-dependent meaning of that word in its more traditional contexts from which it was derived.[38] And where is this strategy of the chimera most obvious? In Spinoza's writing on the name of God where he creates "the impossible equation" God, or Nature.[39]

Yet this equivocation, this directing of attention away from the perpetual looping of analogical selection, is often confused with something altogether less interesting. For it is common for Spinoza to be held up as the progenitor of a "clear-thinking," positivist style atheism.[40] In part this is due to Spinoza's criticisms of religion especially as found in his *Theological-Political Treatise*, but the real persistence of this reading of Spinoza is a fundamental distrust in Spinoza's words by modern readers. This strategy of reading Spinoza is traced by Nancy K. Levene who locates the general thrust of it at work in Leo Strauss's study of Spinoza where Spinoza's real meaning is esoteric. Thus, as modern secular readers, Strauss contends that we are able to read the true meaning of what is hidden in Spinoza's text. Strauss, following Harry Wolfson, holds that at the time of writing Spinoza (like Maimonides) was unable to simply lay out his philosophical system without risking a hostile religious response.[41] If this were true there would be nothing of any real interest in Spinoza. His work would simply be a proto-positivist philosophy of common sense and indeed this is what many crude atheist readings of Spinoza present. Like Thomas Jefferson's construction of a humanist version of the New Testament, these thinkers attempt to scrub Spinoza's thought of all theology and yet when you take God (or nature) out of the *Ethics* you are left with a system that no longer functions and you are left with a politics (*Theological-Political Treatise*) and an ontology (*Ethics*) ultimately unrelated at the level of concept and practice.

Interestingly these esoteric readings run aground because Spinoza does not hide his equivocation of God and Nature. If he were writing secretly to

200   A NON-PHILOSOPHICAL THEORY OF NATURE

those who knew he didn't really believe in God, why would he come right out into the open and write, in perhaps the clearest section of his *Ethics*, "[t]hat eternal and infinite being we call God, *or* Nature, acts from the same necessity from which he exists."[42] For this is actually the productive power of Spinoza's philosophy. Consider that the general formula of Spinoza's ontological scheme can at first appear to be a kind of rationalist reworking of Neoplatonism. From that perspective, the *Ethics* begins with a single substance not unlike the Neoplatonic One and like it this substance must be one since it is that which is in itself and conceived in itself. Therefore substance, like the One, does not require the concept of another thing for it to be formed.[43] This substance is God (or nature) and from God (or nature) attributes can be located, which are what the intellect takes to be the essence of substance.[44] Spinoza holds that there are an infinite number of these attributes, but that the intellect can have knowledge of two: Thought and Extension.[45] There are then also modes that are finite actualities or "affections of a substance, or that which is in another through which it is also conceived."[46] A cursory reading of Spinoza could confuse this schema with a kind of poor Neoplatonic emanationism where substance = the One (not to be confused with the non-philosophical One), attributes = Nous, and the World Soul and its divisions with the modes. But what is productive in Spinoza's thought is the way that emanation is completely changed through the equivocation of God, nature, and substance by way of immanent causality.[47] But what happens through this equivocation is not an overcoding of one term by the other, a kind of secular determination to counter the prevailing theological determination of Spinoza's time, but a complete change in the understanding of both terms.

So, while Levene does not fully deal with Spinoza's Marrano foundation, Strauss and Wolfson aim to overcode the population (of) thought called God with an all-too-easy conception of nature as assumed by the myth of Enlightenment reason. They miss out that both God and nature are changed in Spinoza's thought. As Nancy Levene writes, "From a religious standpoint, God is eternal and nature is made; from a non-religious standpoint, God is made and nature is eternal. Both, to Spinoza, are right."[48] Or as Blayton Polka writes, "*Deus sive natura*. God is (infinitely) natural. But nature is also (infinitely) divine."[49] The minimal difference between these two quotations discloses something important about Spinoza's practice of thought. For Levene's statement it is a question of standpoint, of the thinker thinking the thought, and for Polka's statement it is a question of ontology, what God and Nature *really* are. The reality of Spinoza's practice is that both statements are true, for the standpoint of a thinker thinking a true idea and the being of that idea are the same. This is commonly called parallelism, though as Deleuze reminds us we should be somewhat

MATERIALS FOR A THEORY OF NATURE   201

cautious with this terms since it was not Spinoza's own but appears to come from Leibniz.[50] But Deleuze goes on to say that "Spinoza's doctrine is rightly named 'parallelism,' but this [is] because it excludes any analogy, any eminence, any transcendence. Parallelism, strictly speaking, is to be understood neither from the viewpoint of occasional causes, nor from the viewpoint of ideal causality, but only from the viewpoint of an immanent God and immanent causality."[51] In other words what parallelism really concerns is three forms of identity rather than difference: an identity of order or correspondence between modes of different attributes, identity of connection or equality of principle, and identity of being or ontological unity.[52] In other words, to say that there is a parallelism between the modes of the attributes Thought and Extension is to say that they are the same thing or as Spinoza himself puts it, "So also a mode of extension and the idea of that mode are one and the same thing, but expressed in two ways."[53] One could even take Laruelle's vocabulary and say that in-last-instance they have the same identity and they are distinguished only by the other attributes that their idea involves.[54]

This focus on identity arises in response to a problem. For it would appear that in Spinoza's universe there can be no causal interaction between ideas and bodies. This has to do with the relationship between knowledge of causes and effects, rather than the in-itself reality of those causes and effects. For, with regard to Substance what we can know is actually very little, but we can have knowledge of its essence as expressed in the attributes Thought and Extension. Yet Spinoza states that "[t]he knowledge of an effect depends on, and involves, the knowledge of its cause."[55] Different modes fall under different attributes, that is, they are necessarily conceived in relation to something other than themselves.[56] But since each attribute of a substance must be conceived of through itself, as the essence of the one substance, then a mode falling under one of the attributes cannot affect a mode falling under the other.[57] Yet, just because these finite modes under different attributes cannot causally affect one another does not mean they are unrelated. Instead they share an identity under the three forms Deleuze located and that are discussed earlier. For "[t]he order and connection of ideas is the same as the order and connection of things."[58]

In the *Ethics* this form of thought is only discussed in relation to the modes. But, as Deleuze says, "it is grounded in substance and the attributes of substance."[59] Thus parallelism of modes only makes sense insofar as there is a principle of identity at work. Again, Deleuze captures this point with aplomb:

God produces all things in all attributes at once: he produces them in the same order in each, and so there is a correspondence between modes of

202   A NON-PHILOSOPHICAL THEORY OF NATURE

different attributes. But because attributes are really distinct this correspondence, or identity of order, excludes any causal action of one on another.
Because the attributes are all equal, there is an identity of connection
between modes differing in attribute. Because attributes constitute one
and the same substance, modes that differ in attribute form one and the
same modification. One may in a sense see in this the triad of substance
"descending" into the attributes and communicating itself to the modes.[60]

Thus the doctrine of parallelism is again an expression of the immanence
of God or nature that is present in actualities, though perhaps veiled and
absent as Heidegger suggests.

While we have touched on a number of important elements of Spinoza's
thought and attempted to locate them in their process of directing attention, we have not yet really touched on the importance of this way of
thinking for our non-philosophical and non-theological theory of nature.
Quite simply, Spinoza's theoretical explication of an equivocal immanence
between God and nature breaks naturalism and frees thought to think
nature in the decline of naturalism. In the previous section we credited
Aquinas's apophatic ontology of nature with separating a thought of nature
from a circular materialism and yet we faulted him for creating a kind of
naturalistic fallacy endemic to theological forms of thinking. This was
because Aquinas had to think from what is in order to conceive of God
and what is is varied. It may seem like Spinoza would also be faulted for
a naturalistic fallacy, perhaps more so since he directly equates God and
nature. Yet, Spinoza's equivocation ultimately rejects the pure nature of
naturalism more radically than Aquinas. This is precisely because he separates philosophy and theology from one another in the *Theological-Political
Treatise* while bringing together their ostensible objects God and nature in
the *Ethics*. The confusion lies in the fact that the first separation is ultimately a heuristic one within the practice of thinking itself, while the other
is taken to be an ontological equivocation.

Now, if I am correct that Spinoza radically breaks with a kind of naturalism that posits a simple, closed, and oppressive form of hypostasized
Nature, then the entirety of his theological-political thought depends
on what nature is for him in its chimerical identity as God or nature.
In the *Theological-Political Treatise* we find that Spinoza holds to a position that seems a great deal like Aquinas's. There he writes: "Again, since
nothing can exist or be conceived without God, it is certain that every
single thing in nature involves and expresses the conception of God as far
as its essence and perfection allows, and according the more we come to
understand natural things, the greater and more perfect the knowledge of
God we acquire."[61] He goes on to connect this explicitly to understanding

MATERIALS FOR A THEORY OF NATURE   203

the cause, which is God, through its effects, which is nature or creation, because, "since knowledge of an effect through a cause is simply to know some property of the cause," then "the more we learn about natural things, the more perfectly we come to know the essence of God (which is cause of all things); and thus all our knowledge, that is, our highest good, not only depends on a knowledge of God but consists in it altogether."[62]

There is then an identity of correspondence again between the cause and effect here. Unlike in Aquinas this bears itself out in a political and social thought that moves from the selection of the variance of nature rather than from particulars and this has implications for Spinoza's understanding of natural right. His notion of natural right follows from nature's perfection, which refuses reason demanding anything contrary to Nature and thus demands that "everyone love himself, seek his own advantage, what is really useful to him, want what will really lead a man to greater perfection, and absolutely, that everyone should strive to preserve his own being as far as he can."[63] This is sometimes taken by readers to mean that the strong can do what they will, by their natural power, to the weak who cannot resist by their natural power. Conor Cunningham hyperbolically makes the ridiculous and offensive claim that the equivocation of God, nature, and substance "has therefore enabled Spinoza to rid the world of all substances (and eventually of all substance)" such that "in the world of Spinoza there can be no difference between a Holocaust and an ice-cream."[64] According to this view Spinoza would hold that if one can perpetuate a Holocaust as easily as they produce and eat an ice cream, then there is no reason by right why they should not.

But here we find the Christian theistic theologian Cunningham making the same mistake as atheists like Nadler, whereby he confuses a preexisting concept of nature as the same conception of nature at work in Spinoza. While ultimately this conception is still caught in a dyad with God, it nonetheless suggests a truly radical conception of nature that can be elucidated by putting it in dialogue with the Catholic debate over pure nature. This debate is as complex as it is tedious, but we can distil the doctrine to the idea that says there is some part of God's creation independent of a desire for God. This would mean that there would be some part of nature that was free from the desire for the grace of God and thus sets up a hard dualism between nature and grace, or nature and the supernatural (God), whereas those who stand against pure nature would prefer a softer, grammatical dualism generated by the analogy of being that recognizes only God as true Being. Yet, Spinoza's conception of the relationship between God and nature, even as *natura naturata*, is nothing like "pure nature." Indeed, Spinoza critiques the theology of miracles that contains the seeds that lead to the conceiving of something like "pure nature" within Christian doctrine itself. The

204 A Non-Philosophical Theory of Nature

contemporary theological fixation on pure nature stems in part from the French theological movement *la nouvelle théologie*, the main works of which were written in the early and mid-twentieth century. The argument between *la nouvelle théologie* and the then more accepted neo-Scholasticism of figures like Reginald Garrigou-Lagrange concerned whether or not the theology of the Roman Catholic Church, and thus the theology of Augustine and Aquinas, demanded a stark dualism between nature and grace. For the neo-Scholastics there was a natural ground to meaning with a corresponding natural *telos* that was opposed to the supernatural order of revelation with its own corresponding supernatural *telos*. For *la nouvelle théologie* there was only one order toward one supernatural *telos*. They therefore argued that there is a natural desire for the supernatural; one order and one *telos*. If this was true then it would follow that the neo-Scholastics' conception of the supernatural was merely the natural upon which perfections were added, and this carries with it some questionable Christology and soteriology, for it begs the question of how something can be saved if it is not assumed. And if it is assumed how can it then be said to not have an integral unity, even if unconsummated?[65] The main figure of *la nouvelle théologie* movement with regard to the natural/supernatural debate is clearly Henri de Lubac. His historical studies argue that the concept of pure nature is not to be found in the theologies of Augustine or Aquinas and that they resisted it because of their "profound realism."[66] But such a profound realism is also found at work in Spinoza, more so in his selection of the variance of nature. Thus we find that Spinoza makes his own argument against pure nature in Chapter 6 of his *Theologico-Political Treatise*.

Spinoza's argument here is clear and can be summed up succinctly as follows: because God is the creator of nature, miracles that transcend the natural order would imply an imperfection in God's will and intellect and thus lead to atheism. He writes,

[S]ince the virtue and power of nature is the very virtue and power of God and the laws and rules of nature are the very decrees of God, we must certainly believe that the power of nature is infinite, and its laws so broad as to extend to everything that is also conceived by the divine understanding. For otherwise what are we saying but that God has created a nature so impotent and with laws and rules to feeble that He must continually give it a helping hand, to maintain it and keep it going as He wills; this I certainly consider to be completely unreasonable.[67]

Miracles as normally conceived appear as events whose cause is unknown in nature and from there ascribe a divine origin to them. Spinoza takes us through the logic of this writing, "They also suppose the existence of God

MATERIALS FOR A THEORY OF NATURE   205

is proven by nothing more clearly than from what they perceive as nature failing to follow its natural course. [...] They evidently hold that God is inactive whilst nature follows its normal course and, conversely, that the power of nature and natural causes are superfluous whenever God is active."[68] The position Spinoza is here arguing against is the same understanding, though in a crude mode, that the neo-Scholastics gave to the split between the natural and the supernatural. Spinoza rejects this on the basis of knowledge of nature, as opposed to the theologically orthodox understanding of the revelation of Scripture, but knowledge of nature is taken to be revelatory for questions of metaphysics in the same way that Scripture is taken to be revelatory for questions of true piety (charity and justice) and in this way the two are not opposed.

Consider how Spinoza mutates Hobbes's conception of natural right as explained by Deleuze through his explication of four theses that come out of this mutation. The first is that the law of nature refers to an initial desire and not a state of final perfection. The second thesis is that reason is only secondary to *conatus*, or the will to persevere in one's being. The third thesis states that power or right is primary and unconditional, which is to say prior to reason. The final thesis has two parts: (1) regardless of their powers of reason every person in the state of nature judges what is good, bad, and necessary for their preservation; and (2) no one gives up their natural right due to some recognition of the authority of a wise person, but from either fear of a greater evil or hope of a greater good. The consequence of these theses is that the principle of consent, whether pact or contract, replaces authority as first principle for political philosophy.[69]

Let us return to the second thesis where Deleuze says, "*[No]body is born reasonable*. Reason may perhaps apply and preserve the law of nature, but is in no sense its principle or motive force. Similarly, *nobody is born a citizen*. The civil state may preserve the law of nature, but the state of nature is in itself presocial, precivil. Further still, *nobody is born religious*."[70] Deleuze is here referencing Spinoza's discussion of the state originating in the natural and civil right of individuals by way of pact with sovereign powers found in Chapter 16 of the *Theological-Political Treatise* where we find Spinoza responding to a potential criticism that there is a contradiction between the claim that everyone who is without the use of reason has the sovereign natural right in a state of nature to live by the laws of appetite and the claim that all are responsible before the revealed divine law. Spinoza states that this is true only with respect to humanity's ignorance in the state of nature:

> We can easily deal with this objection simply by examining the state of nature closely. For this is prior to religion both by nature, and in time. No one knows from nature that he is bound by obedience towards God.

206 A NON-PHILOSOPHICAL THEORY OF NATURE

Indeed, he cannot discover this by reasoning either; he can only receive it from a revelation confirmed by miracles. Hence prior to revelation, no one is obliged by divine law, which he simply cannot know. The state of nature is not to be confused with the state of religion, but must be conceived apart from religion and law, and consequently apart from all sin and wrongdoing. This is how we have conceived it, and have confirmed this by the authority of Paul.[71]

This may seem at first as if it were setting up a real separation between something like nature and revelation, but in actuality what Spinoza is doing here is discussing an image presented to human thought of what must be rejected. Indeed, "the state of nature itself shows us what makes it intolerable" insofar as what is fully present there is individual power that ultimately destroys itself.[72] This fabulation is to be contrasted with his repeated statement that "[n]evertheless, no one can doubt how much more beneficial it is for men to live according to laws and the certain dictates of reason, which as I have said aim at nothing but men's true interests."[73]

There are two fundamental points we are to take from this—(1) humanity is not an imperium within an imperium, we are part of nature; and (2) we are able to understand the whole of nature and live in accord with it (also the practical thrust of Spinoza's *Ethics*, the writing of which was interrupted by work on the *Theological-Political Treatise*). This second point may at first appear unclear, but can be elucidated by turning again to Deleuze's reading of Spinoza where he writes,

> There could be only one way of making the state of nature viable: by striving to *organize its encounters*. Whatever body I meet, I seek what is useful. But there is a great difference between seeking what is useful through chance (that is, striving to destroy bodies incompatible with our own) and seeking to organize what is useful (striving to encounter bodies agreeing in nature with us, in relations in which they agree). Only the second type of effort defines *proper or true utility*. [...] There is in Nature neither Good nor Evil, there is no moral opposition, but there is ethical difference.[74]

Such an ethical difference is to be located in the affections that determine a certain *conatus*. Thus the more joyful affects a person experiences and fosters the more ethical they become and the only way to increase joy is through useful encounters or between mutually joyful affects. The more we understand about the world the more we understand how our bodies are compatible with others. In ecological terms, the human species has the natural right, through its natural ability, to drive forward the sixth great spasm of extinction, but it does so at great peril to itself and thus to a decrease in its ability to preserve its being.

MATERIALS FOR A THEORY OF NATURE 207

Ultimately Spinoza will argue that the State is superior to religious institutions in bringing about relationships that accord with the common good (living in accord with nature), for though the Divine Law is higher than the laws of the State, "God has no special kingdom over men except through those who hold power" and "for this reason divine teachings [the practice of justice and charity], whether revealed by natural or prophetic light, necessarily acquire the force of a decree not directly from God, but from those who exercise the right of governing and issue edicts or by their mediation."[75] The State comes to be the site of "true religion," for if justice and charity constitute the thrust of the simplicity of the scriptures and these are fostered within the state more than within the church, then piety is demonstrated in its highest form in the service of the peace and tranquility of the commonwealth, which cannot be preserved if every man is to live simply as an individual.[76] Some have read this as a rejection of religion in favor of the State, but there is another reading that we will follow. In his attempt to separate philosophy and theology, Étienne Balibar points out, Spinoza's turning away from theology and toward immanence may in actuality be a turning toward another, more subtle theology. And so Spinoza finds himself in the awkward position of defending true religion after having set out to defend freedom of thought from theology.[77] One can read as a consequence of his equivocation in the *Ethics* of God or Nature that everything is essentially religious; everything directs its attention toward God or nature. Thus we can read his political thought, and perhaps we should take him at his word here, as theological (though not in terms of dogma) or interior to religion itself. In this philosophy religion is plunged into *conatus* as *conatus* is prior to organized religion (and all organization), for no one is born religious, but religion also names the formation of a relationship between humanity and an indifferent "outside" (nature) that is connected to *conatus*—therefore *it is natural* to be religious even though no one is born so.

Goodchild calls the equivocal immanence of God and Nature cynical in Spinoza, meaning it directs attention toward that which matters most. Thus Spinoza's rejection of pure nature actually gives rise to a politics, rather than taking what is as a sign of the politics that ought to be. Consider that for Spinoza salvation is available to all according to their natural ability (meaning simply to the degree that they actualize their knowledge). Thus for those who are able to think adequate ideas of God they are saved in that knowledge:

> The third kind of knowledge proceeds from an adequate idea of certain attributes of God to an adequate knowledge of the essence of things, and the more we understand things in this way, the more we understand God.

208   A NON-PHILOSOPHICAL THEORY OF NATURE

> Therefore the greatest virtue of the mind, that is, the mind's power, *or* nature, *or* its greatest striving, is to understand things by the third kind of knowledge.[78]

But the ignorant may also find salvation in simple obedience to the requirements of grace and charity demanded of them by religious revelation.[79] In each case, however, the point is that salvation happens in the midst of the earth. The equivocation of God and nature directs attention toward the earth, rather than toward some transcendent beyond, and yet it is able to so without recourse to pure nature. And it does so by making of the earth a form of transcendence produced by what is immanent. For what is taken as revelation of transcendence, religious revelation, is revealed in nature. This is what we will go on to call a non-thetic transcendence or a transcendence that is an effect of immanence, but still expresses the real kernel of transcendence in actuality.

The Paradoxical One of Ismaili Islam and Non-Theological Nature

The choice of Aquinas and Spinoza as paradigmatic of particular practices of thinking through the dyad of God and nature is useful in part because both thinkers are so systematic in their thought. Yet, this systematic character can be faulted for confusing the identity of nature with the character of a system itself. A philosophical system does not mirror nature, nor does an ecosystem. The system and nature are two, even if some aspects or "occasions" of nature are produced in the ecosystem. They become confused in both Spinoza and Aquinas because they are ultimately thinking God and nature in terms of Being, rather than in-One. This problem is analogous to the one that Laruelle locates in materialism:

> In the last instance, it continues to subordinate matter to the ultimate possible form of the *logos* (the *logos* or Idea of matter as such), rather than subordinating the *logos* of matter to matter, thereby engaging a genuinely dispersive becoming-real of ideality instead of a continuous becoming-ideal of the real. Thus, in order to remain faithful to its original inspiration and secure a definitive victory over idealism, materialism should first consent to partially eliminate itself as category and statement—to subordinate its materialist statements to a process of utterance that would be material, relative, or hyletic in itself, then stop conceiving of this utterance as an ideal and relative process. *The decline of materialism in the name of matter, and of matter as hyle in the name of the real.*[80]

MATERIALS FOR A THEORY OF NATURE   209

And so here we must begin to think the radical decline of naturalism in the name of nature and of the dyad of God and nature as natural in the name of the Real. The problem with this can be located by contrasting the systematic *what is* of both Aquinas and Spinoza with a messianic desire for *what is futural* (outside of a simple linear "to come"). What is at work in the dyad of God and nature is a process of absence and veiling. Because nature is in some sense closer to us than our own selves, it is veiled much like the lenses one looks through to see. And it is absent insofar as nature seems to be a universal totality that can never be fully brought to presence. Thus, one creates certain fictions about nature and even Spinoza, the most virulent antianthropomorphic thinker covered here, talks about nature in terms of a human fiction when he talks about the "laws of nature." Nature comes to take the force of law precisely because it is both veiled and absent in ontological thought. It becomes simply what is, even when what is isn't pure nature.

This problem can be brought into sharper focus by looking at the way nature is brought to presence in the other dyad present in the fourfold, that of sky and mortals or what we suggested could also be called the identity of culture. Now, in order to break with the *sékommça* ("it's like that" or "that's how it is") of naturalism, which is to break with the onto-logical cage, we need to think a radical creatural messianism. Something that breaks the cultural and the natural at once. Yet, the two standard options for thinking nature outside the limits of naturalism subsume that messianism into something that already is. I am referring, of course, to the analogical conception and the conception from a position of absolute immanence. Both end ultimately by making the messiancity of creatures impossible. For Aquinas the creature is negated in its relationship with the Creator. For Spinoza immanence is made absolute in the selection of the whole of nature. Yet this ultimately reduces nature to being and this selec-tion of nature as whole or "One-All" retains a transcendent shadow of a quasi thing.[81] This allows for a certain liberty in nature, but such liberty is always limited as determined by its status as Being or what is. In both instances creatural messianity is impossible because what *is* has already been consummated by the death and resurrection of the historical Jesus or, in less dogmatic terms, what *is* is already good and requires nothing more than right order or an ethical relation. In both cases nature comes to be a name simply for the *sékommça*. What is needed, then, is some way to think a messianism that is totally contra-Nature as hypostasized without lapsing into a pure negation of nature by confusing it with the World.

Ya'qûb al-Sijistânî provides us with a proto-non-theological conception of nature that differs from both Aquinas and Spinoza. Whereas Aquinas

210   A NON-PHILOSOPHICAL THEORY OF NATURE

occludes nature by making creation its proper apophatic name and
Spinoza occludes creation within a subsumption of God and nature into a
One-All substance, al-Sijistânî places nature in the middle of six Creations
that come from the uniqueness of the Creator for whom even essence is
excluded.[82] By placing nature as a creation from a Creator that has neither
Being nor essence, al-Sijistânî frees this nature/creation from being subsumed into Being. These are not seven distinct, linear creations, but seven
cyclical Creations that are contained in each other as expressions of the
manifestation of the unique nature of the One, which begins in creation
and runs down through the created from universe to angels to peoples
to prophets before culminating in the resurrection. In non-philosophical
terms, they are clones of the foreclosed One.

Nature is the third Creation and is treated alone, while in the second,
fourth, and fifth Creations there are always two terms that move from
one to the other. By treating nature alone in this way nature is raised to
the same level as the Intelligence (from which the Angels are given their
power and identity) and the Resurrector for Sijistânî. Sijistânî raises nature
to these levels by locating the real identity of nature as Earth rather than
World:

> Do you not see that the human being, in whom the most subtle quintes-
> sence of the two universes has been concentrated, lives on the earth? His
> subsistence is on the Earth. His return is a returning to the Earth, and his
> resurrection is a resurgence from the Earth. From these premises it follows
> that we have shown that the Earth is not inferior to heaven in dignity and
> merits the presence of the angels, since a great number of potentialities
> [*puissances*] are achieved in the Earth that are in harmony with the angels.
> Understand this.[83]

Christian Jambet sums up al-Sijistânî's conception of nature by saying
that it is not being or *physis*, but that allows for the appearance of every
phenomenal existent.[84] But this allowing for appearance is different than
the invariant World as we saw it in Heidegger and Badiou. For here this
appearance is tied directly to that which has no Being or essence, which
the World enthrones. Nature is then the condition for the appearance of
what is totally different from the World, for what is the messianity of creatures rather than as worldly.

The solution to the impasse between analogical transcendence and
absolute immanence, where there can be no messianity, is then a conception of nature as clone of the Real-One. Instead of trying to conceive of the
relationship between the Absolute—God and (hypostasized) Nature—
which requires then some epistemological organon of selection, foreclose

MATERIALS FOR A THEORY OF NATURE   211

the One to thought as that which is beyond Being and Alterity, but which manifests itself as lived in-Person. In terms whose meaning is more directly understandable, instead of treating *Deus sive natura* as a relation between two terms, treat the equation as itself relative to the radical autonomy of the Real-One. This radicalizes the Spinozist response to the Thomist failure of selection, for instead of selecting the whole of nature as the best way to think God, the non-theologian selects the whole of the dyad of God and nature in order to think the cloned One.

Jambet shows how this conception of nature is dependent on being thought from the paradoxical One of Ismaili Islam. Thus al-Sijistânî's conception is helpful for our theory of nature because it moves thinking from the perspective of Being to a thinking in-One. Ismaili Islam, whether or not Laruelle himself knows this, appears as a kind of proto-non-philosophical conception of the One, though couched in terms of a radical transcendence rather than a radical immanence. Yet the formalism of the One, as discussed in chapter 6, can be seen "in-person" so to speak when we think it alongside of the Ismaili experience of liberty, which was actually lived in the proclamation of the time of Resurrection (*qiyâmat*) in the twelfth century, after the collapse of their Fatimid caliphate that ruled over the Islamic world.[85] The story of the Ismailis of Alamût is fascinating and should be of interest to anyone interested in messianism, but the historical details are not of particular interest here.[86] Rather, it is the relationship between this messianic act and the One that is important. Jambet explains this necessary connection and mutation of the Neoplatonic One and the messianic act in Ismaili thought writing:

It is no less suggestive to note, in these two cases, the following philosophical fact: in order to problematize a messianic event, whether it be a fervent premonition or already experienced, it is necessary to interrogate the nature of the One, the nature of the procession of existants, and also to interpret the messianic event according to the laws of engendering the multiple from the One. Why was this theoretical schema so necessary?

It seems to us that there are two simple enough reasons for this. First and foremost, the neo-Platonic schema of the One and the multiple permits the One to be situated beyond any connection with the multiple wherein it would be totalized or counted as one. The One is thought beyond the unified totality of its emanations in the multiple. On the other hand, freed from any link with the totality of the existant, and situated beyond Being, the One can signify pure spontaneity, a liberty with no foundation other than itself. In this way, the sudden messianic appearance of the Resurrector will be founded in the creative liberty of the originary One; thus, in the necessary reign of the existent, the non-Being that results from the excess of the One will be able to mark out its trail of light.

212 A NON-PHILOSOPHICAL THEORY OF NATURE

But, conversely, this creative spontaneity will also explain the creation of the existent, the ordained and hierarchized formation of universes. Just as much as with the unjustified liberty, the One will be able to justify the procession of the intelligible and sensible, and the gradation of the spiritual and bodily worlds. Avoiding dualism, all while thinking the duality between the One and the order of Being which it interrupts; conceiving, on the other hand, of the unity of order and creative spontaneity—all while preserving the dualist sentiment—without which the experience of messianic liberty was impossible: this is what neo-Platonic thought offered to the Ismaili.[87]

In short, the One allows the Ismaili to think the pure formalism of the Real—there is the non-thetic transcendence found in the negation of Being, interrupting the order of Being and beings, and the immanence of (non-)One or the existant that is beyond any totality, that is pure fissure itself.

The immanental aspect of the Real-One, which is carried in each One, simply cannot be reduced to a totality, to some kind of idea of number. It exists without any ground whatsoever, and this is its source of liberty or autonomy from any attempt to capture it within philosophical or theological structures:

> The Ismaili experience of liberty is not the discovery of the autonomy of consciousness or the political rights of the individual. It is the feeling of a different and powerful idea: liberty is not a moment of Being, and it is even less a piece in the game of the existent. Liberty is not an attribute, but rather a subjective affirmation without foundation. Liberty is not a multiple effect of the One, but it can be nothing but the One, disconnected from whatever network of constraints it engenders or by which, on the contrary, it would come to be seized. Liberty is the experience of this non-Being of the One, through which the One inscribes itself in the universe of both Being and beings as pure alterity.[88]

This reveals something important about the identity of immanence: immanence has no ground and is thus, in some real sense, the other to any form of thinking that searches for a transcendental or absolute ground from which to think. Immanence itself is fissured, it itself is the Real-One and thus every real thing is in-immanence and is immanence-in-the-last-identity. Not as subsumed into immanence as absolute substance, but as lived. This Ismaili One of absolute liberty as already-cloned from the Real-One will provide the necessary material for thinking a non-theological nature from this radical immanence.

What are the consequences of this choice? What does this choice do? And how, finally, does it think a messianism of nature?

MATERIALS FOR A THEORY OF NATURE

213

Nature is no longer an object of knowledge nor is it the object of knowledge that comes to know itself, but itself is the inconsistent condition for any such dialectic. As inconsistent it can surprise the creatural subjects of nature, rather than providing their *sékommça* cage. As such it is itself outside of any of this dialectic as radical immanence underlying the transcendental dialectic. Nature is the identified with the Earth rather than with the World. As al-Sijistânî says, nature does not change state. Even if its parts were to be annihilated, it would still remain as nature, as the condition for the appearance of the messiah as divine potential.[89] Worlds may pass away, but their appearance and passing away depends on the Earth. Even when the Earth does pass away, nature as such will remain as *already*-inherent and *already*-manifest.

This divine potential is ultimately an antinaturalism, one that can be explored and explained through a somewhat unlikely detour of the French Maoist angelology of Guy Lardreau and Christian Jambet. For the figure of the Angel in their work is an attempt to negate this unacknowledged and hallucinatory thinking of nature as bound to law through a necessity of arrival or advent of the Angel. The Angel is a pure negative name. Lardreau writes in the introduction of their *L'Ange*:

It is necessary that the Angel come. And so that he comes, being invisible, he must have been visible in his works, he must have been announced in history, he must have been there, not two objects of desire, that is where the Fathers were lost, but two desires. Or rather, a desire, that is to say a sexual desire, and a desire that has nothing to do with sex, not even the desire for God: rebellion. On the one hand pleasure, jouissance, and on the other not even beatitude. Something still unnamed, that we have called desire under the pressure of language, which we must force into delivering a name to us. But the Angel is anonymous, or polynymous. We only call it that by way of negative metaphors. That's how pseudo-Dionysius wants to speak about that which is God. Negative theology. Speaking about the world before the break from which it will be born, we can say nothing except from the negative. I do not see how else to hold on to the hope of revolution.[90]

L'Ange is partly a polemic against the "revolutionaries of desire" (they discuss very briefly Lyotard and Deleuze and Guattari) in ways that prefigure the now popular criticisms of Slavoj Žižek and Malcolm Bull. Namely, the revolution of desire is fully compatible with capitalism; it doesn't overthrow the Master but replaces him with a new form of the Master.[91] I'm not interested in either responding to this critique or in defending it; in fact, it often seems to me that *L'Ange* suffers from a certain inchoate rage directed at thinkers with whom they share a minimal difference. What is

214   A Non-Philosophical Theory of Nature

interesting to me is how this antagonism toward Lyotard and Deleuze and Guattari manifests as an *antinaturalism* in Jambet and Lardreau.

For Jambet and Lardreau naturalism and antinaturalism is the difference between two different forms of revolution. This dualistic theory of revolution is explored in the central chapter of *L'Ange*, written by Lardreau, entitled "Lin Piao comme volonté et représentation" [Lin Biao as Will and Representation]. There they posit yet another gnostic separation, this time between ideological revolution and the absolute revolt of cultural revolution.[92] This dualism isn't completely foreign to more familiar and popular forms of contemporary Marxism, like that found in Antonio Negri who traces revolutions in terms of the difference between constituted power and constitutive power. Like Negri in this respect Lardreau and Jambet are concerned with uncovering how it is that pure revolt against the Master behind every master, which is cultural revolution, becomes ideological revolution, a form of revolution that merely makes possible a new master as it is tied directly to historical processes such as a new dominant mode of production.[93] In this chapter Lardreau undertakes an empirical case study of this dualistic struggle between the different forms of revolution by locating a form of cultural revolution in the early irruption of Christian ascetic monasticism and its accommodation within the church. This early monastic movement is a form of cultural revolution just as the Great Proletarian Cultural Revolution of Mao's China is one.[94] This was a form of life that, even if it was called Christian, had nothing to do with the worldly church of institutional Christianity that helped to found the "institution" of Europe. Instead, as a form of cultural revolution, it "presented itself as an anti-culture, a calculated inversion, systematic, of all the values of this world."[95] In fact Lardreau locates three essential themes of cultural revolution as an extreme path of "struggle alone": "the radical rejection of work, the hatred of the body, and the refusal of sexual difference—certainly not as a production of *one* indifferent sex or of *n* sexes [...] but as the abolition of sex itself."[96]

Thus cultural revolution is "totally *contra-nature*."[97] This has two different but connected meanings. First, it may mean the rejection of the idea that what *is* simply is. This is the form of nature that we located already at work in both the Christian theological tradition and contemporary naturalism. It is nature as the *sékommça*; nature as the "it's like that."[98] The second meaning has to do with death. The hatred of the fleshly body and the desire (for, whatever Jambet and Lardreau say, this is a desire even if it is, like Job, cried forth as a protest) for the subtle or spiritual body can give birth to the messianic act (what we may name as the coming of Christos Angelos or the Future Christ).[99] As Jambet writes later in his career when he locates the messianic act of the Ismaili community of the

MATERIALS FOR A THEORY OF NATURE   215

Alamut mountain fortress in their overturning of the Law in the name of
a liberty found in living the higher life of contemplating the divine: "The
abolition of the law means we replace it as series of distinct obligations by
a single one, which is that of the sabbath."[100] In other words, one lives the
life of one divine rather than the life of survival.

Now in part we find a hatred in Jambet and Lardreau for nature
because they too confuse nature with the World. Such hatred, Laruelle
says, ends in "so many vicious circles and tendencies" that "mistake the
whole of the phenomenon" for "the heretical struggle is not born from
terror or the specular-whole, which it practically undoes, it is born from
the being-separate of man that is in-Man."[101] Yet, more radically than
Spinoza's offer of salvation as reconciliation with an indifferent nature,
Laruelle suggests that salvation comes by way of gnosis. Gnosis knows that
"the divine creation—the World—is a failure," but also that "the necessity
for salvation is universal."[102] If the choice is really between the authority
of the World (as Nature) or an arbitrary but absolute will of the people (as
contra-nature) captured or manifested in State power become barbaric,
then there is no real choice. In each instance the human, or Man-in-Man
in Laruelle's terminology, is turned into a subject (which is the (non-)One),
but its real identity, which produces this subject as the inconsistent immanence of the One, is obscured within an idea of humanity given from an
authority. And in each case we never stop rebelling, human beings nevertheless rebel, Man is in-struggle. Laruelle puts it this way, "*There is revolt
rather than only evil*; nearly everywhere and always people do not cease
to kill but they also rebel against the most violent powers as the most
gentle."[103]

In Laruelle's work the name of the subject who is in-struggle is Christ.
For Laruelle's non-philosophy the subject is produced as an effect of
the radical immanence of Man or the Human. Thus in this case the
Christ-subject may be the masses. Laruelle recognizes this when he writes
(as we've already quoted in chapter 7), "The Future Christ rather signifies
that each man is a Christ-organon, that is to say, of course, the Messiah,
but simple and unique once each time. This is a minimal Christianity.
We the Without-religion, the Without-church, the heretics of the future,
we are, each-and-everyone, a Christ or Messiah."[104] It is in the positive
religions that Christ is misunderstood: "Christianity is the limit, the whole
content of which is a misinterpretation of Christ"[105] A student of Laruelle's,
Gilles Grelet, goes further and connects Christ with Jambet and Lardreau's
Angel, by separating a "marshmallow" Christ ("The marshmallow offers
the perfect image of relation between the 'fundamentally Christian' West
and Christ, since we know that the soft and very sweet candy does not, in
fact, contain any marsh mellow.") from the *Christos Angelos* ("Angel of all

216   A Non-Philosophical Theory of Nature

the angels, the Gnostic Christ is the Envoy charged with delivering men from their enslavement in this world by liberating in them the knowledge of their origin and the means of getting back to the place from which they have been exiled: *Christos Angelos* frees by the knowledge which gives men the means of rebellion that they are, against all humility, fundamentally driven by.").[106] So in order to understand Christ as generic subject we must understand him as radically separate from Christianity. What better way to do considering Christ from the perspective of a Gnostic-Islamic Christology as present in Ismaili thought?[107]

Non-Philosophy is a practice of liberty from philosophy, from the structure of the World, and not an account of foundation and is thus messianic. By taking the relatively transcendent pole of the fourfold and negating the veiling and absence of the dyad of God or nature, we are able to free the earth from unmessianic divinities. This choice then gives us a conception of nature that unifies a scientific stance toward nature as the One of what appears and the condition of that appearing and an ancient philosophical problem of nature that has all too often ended in a conception of an over-determining nature. Nature, in this middle place between the One and the Future Christ as Resurrector, does not provide any of the usual limitations to human and creatural liberty. Nothing in this conception is "unnatural," for nature is itself perverse here. As the condition for the appearance of messianity of the human and other creatures it stands against what simply is, against the *sékommça*. And it is ultimately here, when nature can be turned against the natural, that we see the unified theory of messianism and nature. Yes, let everyone say with Jambet and Lardreau "let the Angel come!" but understand that the Angel can only come to the earth; it can only overturn the World by overturning the absoluteness of both Being and Alterity. For the earth, like the Angel, has no Master and is everywhere and always already in revolt.

Conclusion

Theory of Nature

A Tripartite Theory of Nature in-One

The ideas sketched in the chapters in this book are simple materials. Understood under the immanental ecology put forth in chapter 9, aspects may be extracted from their ecosystem and put into a relationship with other materials. By way of a conclusion I will now present a theory of the identity of nature constructed from these materials. I remind the reader of the ending that we already gave away in the introduction, for this theory of nature understands the creatural as subject of nature, the chimera of God or Nature as non-thetic transcendence of nature, and the One as radical immanence of nature. In that same introduction I claimed that this theory would come to have a determinate meaning by the end of the book and indeed what has thus far been discussed has been necessary for the production of this theory. For this theory of nature has grown out of the ecosystems (of) thought studied in the last chapter by way of a unified theory of philosophical theology and ecology (which is, of course, itself part of an ecosystem (of) thought as well) developed in part III. This unified theory would not have been possible without the reconception of the division of labor between philosophy and science that Laruelle constructed in his non-philosophy as explained in part II, especially chapter 6. So, in this conclusion I will simply act as an ecologist of thought. Ecologists are able to take people into a field and show them the teeming drama of what seemed hidden before. I will do the same now for this theory of nature.

The Creatural as Subject of Nature

Aquinas begins the second book of his *Summa Contra Gentiles* with a defense of the subject matter of the book: creation. That Aquinas sees the need to give such a defense suggests there would be some who would find the subject of creatures inappropriate and even undignified for a theologian; that is, the subject of creation is undignified for one concerned with divine things. It should be unsurprising that Aquinas connects his discourse on creatures with the overriding concern he has of forming a nonanthropomorphic conception of God, of keeping the veiled under cover. Subsequently it is concerned that metaphysics not erase God's divinity, while still opening up some way for creatures to know God. This is because for Aquinas knowing God bestows on creatures real, though relative, dignity. According to Aquinas, the concern with creation is useful for theology and the instruction of faith for four reasons. First, meditation on God's work of and in creation enables us to reflect and admire God's wisdom.[1] Second, considering creation leads to an admiration of God's power.[2] Third, it incites human souls to love God's goodness.[3] Last, it may endow human beings with a certain likeness to God's perfection insofar as God knows everything through Godself.[4] These reasons clearly separate Aquinas's philosophical theology of nature from the facile natural theology of domestic design arguments like those of William Paley's "Watchmaker analogy." In Aquinas the point of reflecting on nature is not to prove the existence of God, but to disclose the character of God. This is paralleled by a careful, though not labored, reading of Aquinas's famous five proofs for the existence of God. Such a careful reading bears out that Aquinas himself thinks such arguments show the limits of "natural" reason to think of God without revelation. In this vein Eugene F. Rogers writes, "Aquinas sees his charge as a teacher of sacred doctrine in the presence of Aristotle as a charge *to consider nothing God-forsaken.* That is the point, by the way, of the Five Ways: to leave no human or physical motion unrelated to God."[5]

Creation differs from pure naturalism in that what is at work in creation is a particular form of subjectivity. Note again Aquinas's reasons for defending a discussion of nature, for in each case it is a question of a subject relation to what is neither a subject nor an object as pure relational veiling. Stepping away from Aquinas's orthodox vocabulary and syntax we can say that creation is a certain fiction produced by the Real. It is, for that, no less part of reality, but it is such as a kind of mode of that Real. The subjectivity of nature is creation or, in a more precise sense, it is creatural. I use this term to avoid confusion with the act of creation implied in the Scholastic separation of Creator, creation, and creatures.[6]

CONCLUSION   219

For what is meant by creation in Aquinas's defense is not the act of creation, which belongs in his view to God alone as cause, but rather to the whole of the effects of the act of that cause. The creatural is the realm of effect. Standard contemporary forms of what could be termed "Speculative Thomism" (a title that would apply to those who use Aquinas as Jacques Maritain and John Milbank do, as well as, to a lesser extent, Michael Northcott, who was discussed in chapter 4), in an attempt to safeguard Thomism and the humiliation of creatures it produces in its apophaticism of nature, might object that I have made a category error here. Rather than bestowing creatures with subjectivity, they could suggest, all subjectivity, at least all *real* subjectivity, is retained for God. But this is exactly what reveals the unmessianic character of Thomism, its deep connection with secular naturalism and a worship of the *sékommça*. While the messianity of creatures might seem like it could only be taken on faith, it is supported by the loop of analogy that, without any organon of selection, constantly loops upon itself producing this same *sékommça*. It also opens up to the mistakes of positing some pure nature, which is anathema to these same Speculative Thomists, but which they are unable to avoid when they insist on the lack of positive existence for nature as creatures. Creatures, however, are always in the midst of fabulating messianity.

The creatural as subject of nature is expressed in the ecological concepts of niche and biodiversity (chapter 9). For each of these is a fiction, or what Bergson and Deleuze call a fabulation, a kind of story as effect of the One that produces Being and Alterity. It is important that this notion of fabulation not be confused with something unreal or even with Husserl's irreal. But the fabulation is the radical immanence of nature as One, a radical immanence that is the lived bodily existence beyond transcendence of a creature. Yet, importantly, this fabulation is created without there being a reciprocal effect on the radical immanence of that Real-One. For there is a finitude at work in the creatural. No creature is eternal within the organization fabulated into the World, yet each manifests as in-One and thus is nature-in-person. When someone appeals to save creation, to save the creatural, what they are appealing to is the salvation of the subject, of this particular finite fiction that is how the earth or biosphere is lived. Again, such a fiction is real, but is real as an effect of the Real and is thus said to be in-One.

Treating creation as a population (of) thought we are able to remove it from the wider system. This system is entrenched in Being as such, but by removing it from the Greek and Christian ontological system, a goal of non-philosophy in general as discussed in chapter 5, we are able to move beyond the infinite loop of analogy. Here, where the creatural is the name of nature's subjectivity, there is no absolute apophaticism of creation in the name of the Creator's lone claim to Being used to chain and treat the creatural as nothing. Rather analogy can come to function as a kind of energy that is exchanged between and connects various creatures to one another in terms of fabulative likeness. That is, rather than a complete rejection of anthropomorphism, we can begin to think of the relative analogies between human beings and other creatures as an effect of the Real. The complete rejection of anthropomorphism is a commonality between theologians and radical ecotheorists, which even seeps into more popular "Green" discourse. But this actually ends up putting a barrier between human beings and other creatures as it sets up the old division between humans and Nature. Human beings are part of the biosphere, they are natural, and as such there are things in nature that have the form [*morphe*] of human beings. That form will have commonalities with other creatures, while of course having limits as well, but by thinking this kind of relative analogy between creatures, creative of not just anthropomorphism but also arbormorphism or elephamorphism. This relative form of analogy is productive of an inconsistent and open ethic. Such an ethic operates through the direction of attention to the suffering and exile common to all creatures. This ethic of attention needs no other reason than their existence to care for others. By way of a certain productive analogy of beings with the human we can begin to change our attention. But this attention is always guarded from being misdirected from suffering by way of a recognition of its ungrounded character as a fabulation. Thus, when we speak of the bees and think of them democratically (for this is closer to the actual organization of bees than monarchy), we are free to do so in terms of a mass creatural subject that includes the human alongside of the bee, but we do not move from there away from this to a conception of the Real as such.

Chimera of God or Nature as Non-Thetic Transcendence of Nature

I can think of two exquisite ethical examples of such a fabulation discussed earlier, though there are surely a multitude in the liminal space between philosophical theology and literature. The first, likely familiar to our readers, are the numerous stories surrounding St. Francis of Assisi and his relationship with animals. St. Francis was able to call all aspects of creation his brothers and sisters, not as an allegory, but as truly sharing in the likeness of a gift to Christ or the Messiah.[7] The other can be found in Epistle 22 of the tenth-century Islamic esoteric secret society based in Basra, Iraq, and that went under the name of *Ikhwan al-Safa* or the Brethren of Purity

CONCLUSION   221

whom many in Ismaili Islam claimed was the "Hidden Imam" or, again, the Messiah. This epistle, translated into English under the title *The Case of the Animals versus Man Before the King of the Jinn*, details a protest made by the animals against humanity and their violence against them because of their belief "that the animals were their runaway and rebellious slaves."[8] Here we can see that in messianic traditions, albeit ones confusing radical immanence with an absolute transcendence, there is already a movement toward this kind of open and inconsistent ethic of attention toward the *creatural*, rather than the toward closed-species or what could even be called the family, if taken in an enlarged sense.

The Aesopian element at work in theological and philosophical fables like this fascinated Spinoza. For this Aesopian element is a version of the general issue of metamorphosis, which fascinated Spinoza because once he defined substance as being singular it seemed to him that we could then think of bodies changing to such an extent they may not be considered to be of the same nature as they were before. This in turn had important philosophical consequences for thinking through what happens to a human being in the change from infancy to adulthood.[9] This interest in metamorphosis, really in what could be termed the supernatural, goes deeper than a passing literary fancy for Spinoza, for it lies at the heart of the strategy of the chimera discussed in the previous chapter. That strategy of the chimera is of special interest to us when it creates the impossible equation of God, or nature. Remember how Zourabichvili defines the chimera as "whose nature veils [*enveloppe*] an open contradiction" and that which "by its nature, cannot exist," going on to write that "[t]he chimera is not a thing but, if we can put it this way, a non-thing, a *non-nature*."[10]

Here what we find is a conception of nature that is expressed as non-thetic transcendence or what has been expressed as the non(-One) in Laruelle. Though it seems oblique, it refers to "the real kernel of transcendence" or rather transcendence understood as an effect of the radical immanence of the One.[11] As such it is a transcendence that is produced as an effect of transcendence and yet gives itself nonreflexively as the support for philosophical and theological thinking. A thetic statement is tautologous, like $A = A$, and as such fuses the idea of the posited and positing. Paradoxically, then, even a thetic statement is a kind of chimera, an impossible equation like God = nature. The power of a thetic statement lies in the fact that they locate an identity outside of relativity and so do not take into account the directionality of thought. Consider that if a river were to suddenly reverse course, like the Chicago River was forced to do by civil engineering in 1887, it does not cease being a river even though that change in course radically alters its ecological make-up. What is non-thetic about this transcendence is that it is relationally determined while also transcending those relations. By locating a non-thetic form of transcendence in God or nature we show that this equivocation, this chimera, is constructed or fabulated even as it is constitutive of thought, as discussed in chapter 7.

Nature, with the capital N, is thus unveiled in this theory of nature through its cynical or chimerical equivocation with God. What is required for thinking is a fiction of ground that is ultimately ungrounded itself. Thus to see Nature as a veiled open contradiction (a hypostasized Nature) is one identifying it as a chimera. This opens up to thinking the remainder produced by this transcendence as a kind of nature that is no longer recognizable on the terms required by philosophy and theology. This view of Nature is no longer a threat, as we saw in chapter 10 it is for Latour and Morton, but disempowered as part of the fabric of thinking that, while transcending thought as such, is always at play within thought and part and parcel of the practice of thinking. Ecology may not need this notion of Nature for its own scientific practice, but it comes to be a certain kind of never-living aspect of philosophy productive of other thoughts. Thus, one may still develop a thinking of Nature, but such a thought is able to show that in all the various philosophical and theological constructions of Nature they are taking something produced as an effect of radical immanence for the absolute. By seeing this as a fiction, albeit one at the foundation of a certain kind of thinking, worldly Nature is freed to be nature separate from the World (as discussed in chapter 11) and thus to be at play with creatural nature rather than overdetermining it.

The One as Radical Immanence of Nature: Or, How Nature Is a Name for the Real

The One of both Ismaili Islam and non-philosophy is not a unified totality. It is not numerical in any reductive sense. Instead, Laruelle calls upon this ancient philosophical and theological transcendental to characterize the Real as prior (he will say "without priority") to Being and Alterity. In the same way the Ismaili called upon the One as a way to think a liberty with no other foundation except for itself. The One is simply radical immanence, the experience of something prior to the World at work in the Real.

When I first embarked on this philosophical research into ecology, begun largely as a conversation between Husserlian phenomenology and a Bergsonian and Deleuzian understanding of the immanence of nature, I began making the claim that the earth is prior to the World. As a rhetorical

CONCLUSION   223

strategy and in relation to the fourfold I still think this phrase works, but
it is more rigorously expressed as the One is prior to the World. The reason
that the earth isn't a proper name for the Real within a unified theory of
philosophical theology and ecology is because the earth is actually part
of a wider relationality for its identity or, in other words, it is an effect
of nature. There is a kind of reversibility at work, as we say in chapters 8
and 9, between the earth and the flows of energy and material that make it
up, to the degree that the earth really is but a name for a certain organization as One. In the same way the oceans and the atmosphere could also
be said to be prior to the World, but only because they too are a certain
organization as One.

So what does it mean to claim the One as the radical immanence of
nature? It is very different from the naturalisms and materialisms of philosophy presented in chapter 2. Rather, the answer to this is found in
thinking together the lived subjective experience of nature as creatural as
well as its non-thetic transcendence as Nature as produced by this radical
immanence. For the radical immanence of nature as One unifies these two
modes of nature. As a creature I experience nature in-person as my body.
This is true of the lived experience of my body (of which I see no reason to
separate mind, since body is here taken in the sense of an effect rather than
a reductive materialist definition). For as a creatural body I am a certain
ecological organization of matter and energy. I produce in an ecological
sense. Yet, there is also a transcendent element at work in my experience
of this body in the sense that there is something separate from me, transcending in a relative sense, that I may also call nature. I think of the way
an affliction, genetically coded, like gout, may suddenly come upon my
body and change my bodily relations, physical and mental, with the rest of
creatural nature. And yet, in the immediacy of practice and experience, I
don't take this transcending ailment as a parameter set by Nature. That is
a secondary move and even a poor fabulation. Rather, what I can fabulate
as a creature, like Spinoza, is a cynical equivocation that locates within this
same transcending Nature the site of the appearance of the Messiah or a
great number of potentialities achieved on this body. So, within creatural
nature there are others who offer me help. Still others who create medicines
or research into the phenomenon and provide new knowledge about what
to do. So in the midst of this Nature, already disempowered for us now,
we see the appearance of love and charity. We may also see violence, but
such violence is always struggled against (as has already been discussed in
chapters 7 and 9).

Thus the Unicity of nature as One is not something that shall be
thought, if by this we aim to circumscribe and affect nature as One. For
that is impossible and only hallucinates either nature as subject or Nature

224   A NON-PHILOSOPHICAL THEORY OF NATURE

as non-thetic transcendence in the place of radical immanence. Rather this radical immanence of nature is known simply as the unconcealed prior to knowledge. It can't be known philosophically or theologically, but only through a kind of faithfulness as struggle. Is this the same faithfulness one often finds in non-standard naturalists (I'm thinking of those forms of naturalism that aren't simply reductive, as in Deleuze and Guattari, Bergson, and Whitehead) required by Nietzsche's Zarathustra when he cried, "I beseech you, my brothers, *remain faithful to the earth*, and do not believe those who speak to you of otherworldly hopes!"[12] Perhaps, but only if the "otherworldy hopes" is understood as every World that erases the radical immanence of creatural life and death, that erases the radical immanence of an ecosystem. In fact, the entire context of Nietzsche's fidelity must change. For his fidelity to the earth is a fidelity to the meaning that it *shall* come to have after the Overman comes. Rather, fidelity to the Oneness of nature in this case merely points to it being already unveiled and already present as radically autonomous, even while thought itself is unable to think this on the terms of thought. We may instead say that this non-theological faithfulness mirrors the Ismaili faithfulness to the One: "There is a merging of radical apophaticism and the doctrine of epiphany here."[13]

The question of naming the Real as nature thus speaks to a kind of fidelity and even submission, though not to the *sékommça* of Nature or even an already decided Real. The fidelity, however, is neither to some Other as the Real. Nor of course is it to the self. Rather it is fidelity to the revealed of the unrevealable, that which is revealed but which thought is incapable of thinking as something it can reveal. Naṣīr al-Dīn Ṭūsī, who developed a *summa* of Ismaili thought by building on the work of al-Sijistânî, explains the meaning of the title "infidel" or *kafir* in Arabic, "He is called *kafir* because the word '*kufr*' means 'to cover up', meaning that he has concealed from himself what he cannot know."[14] In other words, for the Ismaili as for the non-philosopher, the act of infidelity is to cover up that upon which thought runs aground, it is to stop thinking at the limits of thought, even to simply accept the absolute or divinity. This is the impetus for Ismaili messianism surrounding their hope in the Resurrector, which ultimately is the name of the paradoxical One. Jambet explains that

> it is impossible to name the paradoxical One. This impossibility immediately implies the necessity to name: that which we can say, we must say. That is why the Resurrector's name offers itself to the naming of what is above and beyond the names. As the Real is indescribable and not demonstrable, it is necessary to grasp it, not in itself as such, but in the manifestation of the primordial origination of the human locus of the Divine Imperative, the Resurrector.[15]

CONCLUSION   225

The one who is faithful names from within the midst of the cage naturalism attempts to put around creatures with its *sékommça*. It does so by
merging the radical apophaticism productive of creatural subjectivity with
the epiphany of radical immanence productive of a chimerical transcendence. One is faithful when one submits to the One, which is a promiscu-
ous fidelity. Again following the Ismaili this is not a legalistic submission,
as if here we must submit to "laws of nature" and the limits it places on
human beings.[16] Rather it is a submission to nature as name of the Real,
not as simple Power, but as knowledge and recognition of the freedom
inherent in radical immanence and productive of subjectivity. This allows
us to create a better fabulation of Nature and its relation to creatural subjects, a fabulation that is able to find a joy in creating a different way of
living as creatures, respecting what may be called limits, but which turns
them into something other than negations. We turn nature into a first
name of the Real because this changes a hallucinatory, philosophical, and
theological Nature that loves to hide into a nature that, while foreclosed
to thought, reveals itself in our experience of the everyday. Ontology and
ethics are not treated separately here, for this form of nature as the Real,
as a radical immanence that we ourselves are the lived subjects of, requires
a kind of submission in the form of a certain kind of knowledge, a gnosis
that is salvation from the fabled war between humanity and nature. For
both are rootless. Both are abstract. Both are exiled such that home itself is
always something stranger than it seems.

Notes

INTRODUCTION

1. François Laruelle, *Introduction au non-marxisme* (Paris: PUF, 2002), p. 40.

I NATURE IS NOT HIDDEN BUT PERVERSE

1. François Laruelle, *Future Christ: A Lesson in Heresy*, trans. Anthony Paul Smith (London and New York: Continuum, 2010), p. xxvi.
2. For a history of the various meanings this phrase has taken on from its inception in Heraclitus to Heidegger's ontology, see Pierre Hadot, *The Veil of Isis: An Essay on the History of the Idea of Nature*, trans. Michael Chase (London and Cambridge, MA: The Belknap Press of Harvard University Press, 2006).
3. Ibid., pp. 7–14, 101–137, 155–229.
4. Philip Goodchild calls this a clash between economy and ecology. See Philip Goodchild, "Oil and Debt—the Collision between Ecology and Economy," *Situation Analysis*, 2 (Spring 2003): 5–18.
5. Timothy Morton makes this point well when writing, "We can no longer have that reassuringly trivial conversation about the weather with someone in the street, as a way to break the ice or pass the time. The conversation either trails off into a disturbingly meaningful silence, or someone mentions global warming. The weather no longer exists as a neutral-seeming background against which events take place. When weather becomes climate—when it enters the realms of science and history—it can no longer be a staged set" (*The Ecological Thought* [London and Cambridge, MA: Harvard University Press, 2010], p. 28).
6. St. Thomas Aquinas, *Summa Contra Gentiles* trans. A. C. Pegis, James F. Anderson, Vernon J. Bourke, and Charles J. O'Neil, 4 vols. (Notre Dame, IN: University of Notre Dame Press, 1975), II/4.1. [Hereafter SCG. Citations refer to chapter and section. Volumes will be indicated by roman numerals.]
7. Benedict de Spinoza, *Theological-Political Treatise*, trans. Michael Silverthorne and Jonathan Israel (Cambridge: Cambridge University Press, 2007), p. 10.

228 NOTES

8. Ibid., pp. 10, 13. Nancy K. Levene argues persuasively that Spinoza transforms and expands the meaning of revelation to something universal, rather than something held within a particular religion. See her *Spinoza's Revelation: Religion, Democracy, and Reason* (Cambridge: Cambridge University Press, 2004).

9. Within the scope of this work, specifically in the chapters collected under part I, the majority of the theology dealt with will be Christian, largely Roman Catholic, and while there are important reasons to consider Islamic, Jewish, or other forms of theology at greater length, I nonetheless limit myself and consider this a necessary limitation due to time and space. However, Islamic theology comes to have a privileged place in part IV.

2   ECOLOGY AND THOUGHT

1. This understanding of ecosystem is faithful to the mature formulation by A. G. Tansley in 1935. See A. G. Tansley, "The Use and Abuse of Vegetational Concepts and Terms," *Ecology* 16.3 (1935): 299–303. See also Christian Lévêque, *Ecology: From Ecosystem to Biosphere* (Plymouth, UK: Science Publisher, Inc.), pp. 25–27. I am deeply indebted to my colleague and friend Prof. Liam Heneghan of DePaul University's Institute for Nature and Culture for the concept of the "never-living" and his help in understanding the concept of ecosystem more fully.

2. For one account of the history of the development of the ecosystem concept, see Lévêque, *Ecology*, pp. 15–35. For a longer, more explicitly historical account, see Frank Benjamin Golley, *A History of the Ecosystem Concept in Ecology: More Than the Sum of the Parts* (New Haven and London: Yale University Press, 1993). I will return to the ecosystem explicitly in part III.

3. Lévêque, *Ecology*, p. 27.

4. Daniel B. Botkin, *Discordant Harmonies: A New Ecology for the Twenty-First Century* (Oxford: Oxford University Press, 1990), p. 25.

5. We use the term "image" throughout this section to refer to what Botkin calls at different times models, images, and concepts. By consistently referring to these as images or images of thought we mean to create a connection between these images as dogmatically determinate on the practice of ecological science in a similar way as the dogmatic image of thought was shown to determine philosophy by Gilles Deleuze in his *Difference and Repetition*, trans. Paul Patton (New York: Columbia University Press, 1995), Chapter 3.

6. Botkin provides more historical detail on each image than I do and I refer the reader to his book for the historical specificities of each image. For a fuller picture of the ideas and attitudes toward nature, but without the connection to ecological practices, the reader should also consult Peter Coates, *Nature: Western Attitudes since Ancient Times* (Cambridge: Polity Press, 1998).

7. Botkin, *Discordant Harmonies*, pp. 75–89.

NOTES   229

8. Coates does discuss some theological themes within the specificity of the
    attitude toward nature in the Middle Ages. This, however, is limited in that
    Coates does not connect these themes to actual scientific practices. See Coates,
    Nature, pp. 40–66.
9. Botkin, *Discordant Harmonies*, p. 82.
10. Ibid., p. 87.
11. Ibid., pp. 91–99.
12. Ibid., p. 93.
13. Ibid., p. 92; emphasis, as always unless noted, in original.
14. Ibid., p. 98.
15. Ibid., p. 99.
16. Ibid., pp. 101–110.
17. Ibid., p. 103.
18. Ibid.
19. Ibid., p. 109.
20. Cf. Ibid., p. 108.
21. Ibid., pp. 113–131, 133–151.
22. Ibid., p. 120.
23. Cf. Ibid., pp. 125–127.
24. Ibid., p. 136.
25. Ibid., pp. 146–147.
26. Ibid., p. 151.
27. Ibid., p. 110.
28. See Alain Badiou, *The Century*, trans. Alberto Toscano (Cambridge: Polity,
    2007), pp. 48–57. There he suggests that this passion is what differentiates
    the twentieth century from preceding ones, a claim that would suggest my
    drawing on this notion to explain images of thought from prior centuries is
    misplaced, but his own claim is undercut by calling on forebearers for this
    passion such as Hegel.
29. Ecology has taken great pains to legitimate its scientific character, though
    against doubts cast upon it by the wider scientific community because of its
    ties with political issues. The question of the relationship between environ-
    mentalism or political and scientific ecology is an interesting one with regards
    to this legitimation crisis. See Lévêque, *Ecology*, pp. 3–4, 8–9.

3   PHILOSOPHY AND ECOLOGY

1. Some anecdotal evidence for this majority position of ethics and aesthetics can
    be had by comparing the amount of Google hits one gets, at least in August
    of 2009, for "environmental ethics" (about 1,180,000) and "environmental
    aesthetics" (about 26,900) compared to "ecological metaphysics" (241) and
    "metaphysics of ecology" (9). (Out of curiosity I reran this search in June of
    2011 as I was preparing to submit the final version of my thesis and found that

230   NOTES

all had increased by nearly double. "Environmental ethics" then had about
2,250,000 hits, followed by about 64,800 for "environmental aesthetics,"
while "ecological metaphysics" had about 468 and "metaphysics of ecology"
jumped to about 336. As of December 2012 these numbers had increased
again, though not as dramatically.) Both environmental ethics and environmental aesthetics also have entries in the major encyclopedias of philosophy,
whereas ecological metaphysics does not. Further to this anecdotal evidence
there is a major journal dedicated to environmental ethics (*Environmental
Ethics*), but there is a complete absence of a journal that focuses on the metaphysics of ecology. Even the more far-reaching *Journal of Environmental
Philosophy* tends to focus on ethics and aesthetics (largely from a phenomenological perspective), while metaphysics figures very marginally.

2. Paul W. Taylor, *Respect for Nature: A Theory of Environmental Ethics* (Princeton,
    NJ: Princeton University Press, 1986), pp. 50–51.
3. Ibid., p. 51.
4. Ibid.
5. See Arne Naess, "Spinoza and Ecology," *Philosophia* 7:1 (March 1977): 45–54.
6. Arne Naess, *Ecology, Community and Lifestyle: Outline of an Ecosophy*, trans.
    David Rothenberg (Cambridge: Cambridge University Press, 1989), pp. 66, 67.
7. See Ibid., pp. 171–182.
8. Ibid., p. 36.
9. Ibid., p. 39.
10. Ibid.
11. Ibid., p. 34.
12. Ibid., p. 27.
13. Cf. Ibid., p. 38.
14. Ibid., pp. 57, 57–63.
15. Ibid., pp. 26–28. Though it isn't as if Naess is wrong here; since writing this
    over two decades ago governments have not appeared to become any wiser.
16. Ibid., p. 41.
17. Ibid., p. 48.
18. See Ray Brassier, *Nihil Unbound: Enlightenment and Extinction* (Basingstoke:
    Palgrave Macmillan, 2007), Chapter 1; and Quentin Meillassoux, *After
    Finitude: An Essay on the Necessity of Contingency*, trans. Ray Brassier
    (London and New York: Continuum, 2008), Chapter 1. It is quite pos-
    sible that Meillassoux's conception of correlationism, the idea that nearly all
    post-Kantian philosophy suffers from an antirealism that holds matter and
    some form of mind necessarily exist together, was developed after reading §46
    of Husserl's *The Crisis of the European Sciences and Transcendental Philosophy:
    An Introduction to Phenomenological Philosophy*, trans. David Carr (Evanston:
    Northwestern University Press, 1970).
19. Husserl, *The Crisis of European Sciences*, p. 5.
20. Ibid., p. 6.
21. For his attack on meaning, see Brassier, *Nihil Unbound*, p. 239.
22. Husserl, *The Crisis of European Sciences*, p. 51.
23. Ibid., p. 50.

NOTES   231

24. Ted Toadvine makes the same argument regarding Nature and environmental sciences with special reference to Merleau-Ponty's philosophy. See Ted Toadvine, *Merleau-Ponty's Philosophy of Nature* (Evanston: Northwestern University Press, 2009).

25. Husserl, *The Crisis of European Sciences*, p. 112.

26. See ibid., p. 59.

27. Eugen Fink, *Sixth Cartesian Mediation: The Idea of a Transcendental Theory of Method*, trans. Ronald Bruzina (Bloomington: Indiana University Press, 1995), p. 144.

28. Husserl, *The Crisis of European Sciences*, p. 135.

29. Ibid., p. 148.

30. Fink, *Sixth Cartesian Mediation*, p. 32.

31. Husserl, *The Crisis of European Sciences*, p. 150.

32. Ibid., p. 175.

33. Ibid., pp. 113–114.

34. Michel Henry, *I Am the Truth: Toward a Philosophy of Christianity*, trans. Susan Emmanuel (Stanford: Stanford University Press, 2003), p. 38.

35. Ibid.

36. Husserl, *The Crisis of European Sciences*, p. 120.

37. Ibid., p. 152.

38. I provide here just a sampling of the texts. Bruce V. Foltz's *Inhabiting the Earth: Heidegger, Environmental Ethics, and the Metaphysics of Nature* (New Jersey: Humanities Books, 1996) provides a study of Heidegger's considerations of the environment and nature. Toadvine's *Merleau-Ponty's Philosophy of Nature* does the same with regard to Merleau-Ponty's philosophy. The edited collection *Eco-Phenomenology: Back to the Earth Itself*, eds. Charles S. Brown and Ted Toadvine (Albany: State University of New York Press, 2003), includes a number of essays on particular phenomenologists, and a few, notably by David Wood and John Llewelyn, on the general notion of an ecophenomenology, but these both fail to engage with the ecosystem concept. There are a number of books that engage with related ecological issues, including the edited collection *Nature's Edge: Boundary Explorations in Ecological Theory and Practice*, eds. Charles S. Brown and Ted Toadvine (Albany: State University of New York Press, 2007), which considers different ideas of boundaries, human/animal, living/dead, but does not include any essays that engage with the ecosystem concept. Brett Buchanan's *Onto-Ethologies: The Animal Environments of Uexküll, Heidegger, Merleau-Ponty, and Deleuze* (Albany: State University of New York Press, 2008) provides an interesting study of the of protoecologist Jacob von Uexküll's notion of the *Umwelt* and the differing forms of influence it had on Heidegger, Merelau-Ponty, and Deleuze, who of course was not a phenomenologist. While this doesn't provide much in the way of an ecological philosophy, it does provide a resource for bridging the gap between animal-philosophy and ecological philosophy. In terms of the structure we're tracing here, the limit of philosophy and science, none of these works moves beyond the phenomenological drive to subsume science into itself. There is, however, Robert Frodemen's *Geo-Logic: Breaking Ground between Philosophy*

232 NOTES

and the Earth Sciences (Albany: State University of New York Press, 2003), which is a work that, while sharing a different theoretical orientation, is in the same spirit as this work. Instead of a unified theory of philosophical theology and ecology Frodemen present a kind of unified theory of phenomenology and geology. While his work is then not of particular use here, it does give witness to the perversity of nature and the infinite task of thinking from nature.

39. John Llewelyn, *The Middle Voice of Ecological Conscience* (Basingstoke: Macmillan, 1991); and *Seeing through God: A Geophenomenology* (Bloomington and Indianapolis: Indiana University Press, 2004).

40. Llewelyn, *Seeing through God*, p. 22.

41. John Mullarkey makes this point about Llewelyn's philosophy writing, "Each and every being has a claim on me on account of the 'that it is' rather than the 'what it is' of each; the ontological rather than the ontic. There need be no qualitative similarity (having consciousness, sentience) between us, and yet there remains an ethical responsibility all the same." John Mullarkey, "A Bellicose Democracy: Bergson on the Open Soul (or Unthinking the Thought of Equality)," in *After the Postsecular and the Postmodern: New Essays in Continental Philosophy of Religion*, eds. Anthony Paul Smith and Daniel Whistler (Newcastle-upon-Tyne: Cambridge Scholars Publishing, 2010), p. 178.

42. Llewelyn, *Seeing through God*, p. 21.

43. On the status and use of the Gaia hypothesis within scientific ecology, see Felix Baerlocher, "The Gaia Hypothesis—A Fruitful Fallacy," *Experientia* 46.3 (1990): 232–238.

44. Meillassoux uses this phrase in his *After Finitude*, saying that contemporary philosophy has forgotten the "great outdoors" and that he aims to mark a return to it. Meillassoux, *After Finitude*, p. 7.

45. See Iain Hamilton Grant, "The 'Eternal and Necessary Bond between Philosophy and Physics': A Repetition of the Difference between the Fichtean and Schellingian Systems of Philosophy," *Angelaki: Journal of the Theoretical Humanities* 10:1 (April 2005): 43–59.

46. Grant, "The 'Eternal and Necessary Bond between Philosophy and Physics,'" 44.

47. Ibid., 46–47.

48. Ibid., 50.

49. For Žižek's criticism of the ideology of ecology and his own attempt to recast the problem of nature alongside of ecology, see "Unbehagen in der Natur," in his *In Defense of Lost Causes* (London and New York: Verso, 2008). For Žižek's indebtedness to Schelling, see his reading in Slavoj Žižek, *The Indivisible Remainder: On Schelling and Related Matters* (London and New York: Verso, 2007); and for an account and criticism of that reading, see Iain Hamilton Grant, "The Insufficiency of Ground: On Žižek's Schellingianism," in *The Truth of Žižek*, eds. Paul Bowman and Richard Stamp (London and New York: Continuum, 2007), pp. 82–98. For an interesting comparative study of Žižek's and Heidegger's thoughts on nature, which argues that they share a vision of a fundamentally not-whole nature, see Michael Lewis, *Heidegger beyond Deconstruction: On Nature* (London and New York: Continuum, 2007), pp. 105–127.

NOTES   233

50. See Žižek, *The Indivisible Remainder*, pp. 189–231.
51. Ibid., p. 208.
52. Iain Hamilton Grant, *Philosophies of Nature after Schelling* (London and New York: Continuum, 2006), p. 194.
53. Grant, *Philosophies of Nature*, pp. 19–21. My understanding of Grant's extensity test, along with his reading of Schelling's philosophy of nature and its limits, is dependent on conversations with Daniel Whistler, who also constructed this block quote. See Daniel Whistler, "Language after Philosophy of Nature: Schelling's Geology of Divine Names," in *After the Postsecular and the Postmodern: New Essays in Continental Philosophy of Religion*, eds. Anthony Paul Smith and Daniel Whistler (Newcastle-upon-Tyne: Cambridge Scholars Publishing, 2010), pp. 335–359.
54. Grant, *Philosophies of Nature*, p. 20.
55. A philosopher that thinks along the same lines as Grant may respond to this saying that we have fallen into a kind of philosophy of organics, like Fichte and the neo-Fichteans that Grant identifies, but, as is clear from our discussion of the formation of the ecosystem concept, the ecosystem concept is not compromised in this neo-Fichtean forgetting of nature, of the never-living.
56. Edward O. Wilson, *The Diversity of Life* (London: Penguin Books, 2001), p. 169.
57. Lévêque, *Ecology: From Ecosystem to Biosphere* (Plymouth, UK: Science Publisher, Inc., 2003), p. 4.
58. Wilson, *The Diversity of Life*, p. 153.
59. "Unlike physics or genetics, for example, ecology has not created a significant construct of organized laws. In this sense, many propositions of ecology, such as the concepts of niche, climax, or even biosphere, cannot be tested in the sense that Popper defines testing. But one of the reasons why we sometimes question the status of ecology as a science is that it is often difficult to eliminate the particular point of view of the observer and to eliminate all value judgements on the object of study" (Lévêque, p. 4).
60. Wilson, *The Diversity of Life*, pp. 153–154.
61. Ibid., p. 154.
62. Ibid., p. 155.

4   THEOLOGY AND ECOLOGY

1. This is a concept I develop in relation to Laruelle's non-philosophy and questions within the philosophy of religion. See my "What Can Be Done with Religion?: Non-Philosophy and the Future of Philosophy of Religion," in *After the Postsecular and the Postmodern: New Essays in Continental Philosophy of Religion*, eds. Anthony Paul Smith and Daniel Whistler (Newcastle-upon-Tyne: Cambridge Scholars Publishing, 2010), pp. 280–298.
2. E. O. Wilson, *The Creation: An Appeal to Save Life on Earth* (New York: W. W. Norton & Company, 2006).

234   NOTES

3. See Martin Heidegger, "The Question Concerning Technology," in *Basic Writings*, ed. and trans. David Farrell Krell (San Francisco: HarperSanFrancisco, 1993), pp. 317–318, 320. Here Heidegger connects the forgetting of Being with the concealment of the Earth in its presenting itself as "coal to be mined and hauled."

4. See Willis Jenkins, *Ecologies of Grace: Environmental Ethics and Christian Theology* (Oxford: Oxford University Press, 2008), pp. 77–92, 153–188.

5. Karl Barth, *Church Dogmatics* III.4, trans. A. T. Mackay, T. H. L. Parker, H. Knight, H. A. Kennedy, and J. Marks, eds. G. W. Bromiley and T. F. Torrance (London and New York: T&T Clark, 2004), p. 349. I'm indebted to Jeremy Ridenour for a discussion about Barth and animals.

6. Michael S. Northcott, *The Environment and Christian Ethics* (Cambridge: Cambridge University Press, 1996), p. 227.

7. This insight is properly a non-philosophical one and comes from John Mullarkey's own non-philosophical study of film-philosophy, where he argues from the insight: "There is always what one might call the 'transcendental choice of film' at work in film-philosophy: by this I mean those (inadvertently) illustrative approaches that use particular films to establish a theoretical paradigm of what film is and how it works. Yet, such approaches already make their selections of particular films or film elements (of plot over sound, or framing over genre, and so on) in the light of the theory of film in question, and are therefore circular" (John Mullarkey, *Refractions of Reality: Philosophy and the Moving Image* [Basingstoke: Palgrave, 2008], p. 5).

8. Northcott, *Environment and Christian Ethics*, p. 83.

9. Ibid., pp. 1–32.

10. Ibid., p. 57.

11. Ibid., p. 61. Here he mentions some of the figures that the theologians we cover under the inflection type are drawn to, such as Lovelock and Fritjof Capra.

12. Ibid., p. 191.

13. Ibid., pp. 83–84.

14. Ibid., p. 57.

15. Pope Benedict XVI, *Encyclical Letter, Caritas in veritate: Charity in Love* (Rome: St. Peter's, 2009), 4; para 51. Available online: http://www.vatican.va/holy_father/benedict_xvi/encyclicals/documents/hf_ben-xvi_enc_20090629_caritas-in-veritate_en.html (accessed June 13, 2010).

16. Sallie McFague, *The Body of God: An Ecological Theology* (London: SCM Press, 1993), p. 29. McFague, unlike Northcott, does not locate a golden age, though she, unlike Ruether, does locate a prior age of "wilderness," resistant to dualisms, that gives her theology the shape of being formed a bit more by the declension narrative than other theologians in the inflection type. See ibid., p. 4.

17. Leonardo Boff, *Cry of the Earth, Cry of the Poor*, trans. Philip Berryman (Maryknoll, NY: Orbis Books, 1997), p. 8.

18. Rosemary Radford Ruether, *Gaia & God: An Ecofeminist Theology of Earth Healing* (London: SCM Press, 1992), p. 3.

NOTES   235

19. Ibid.
20. McFague, *Body of God*, p. 29.
21. Ibid., p. 3.
22. Ibid.
23. Ruether, *Gaia & God*, p. 39.
24. Both Ruether and McFague make reference to the Gaia hypothesis. McFague in relation to her understanding of ecology as concerned with "unity" (McFague, *Body of God*, pp. 29–30) and Ruether introduces Lovelock very early on (Ruether, *Gaia & God*, p. 4).
25. Leonardo Boff, *Ecology and Liberation: A New Paradigm*, trans. John Cumming (Maryknoll, NY: Orbis Books, 1995), p. 10.
26. Ruether, *Gaia & God*, pp. 1–2.
27. Boff, *Ecology and Liberation*, pp. 9–12.
28. See McFague, *Body of God*, pp. 27–63; Ruether, *Gaia & God*, pp. 32–58; Boff, *Cry of the Earth*, pp. 140–157.
29. Boff, *Ecology and Liberation*, pp. 43–45.
30. Boff, *Cry of the Earth*, p. 10.

5 Theory of the Philosophical Decision

1. John Mullarkey, *Post-Continental Philosophy: An Outline* (New York and London: Continuum 2006), p. 148.
2. Brian Walker and David Salt, *Resilience Thinking: Sustaining Ecosystems and People in a Changing World* (Washington and London: Island Press, 2006), p. 145.
3. Cf. François Laruelle, *En tant qu'Un: La « non-philosophie » expliquée aux philosophes* (Paris: Aubier, 1991), p. 253.
4. In one of his latest works Laruelle shows that he ascribes to "the communist hypothesis," as Alain Badiou has described it, when he marks an equivalency between his long-standing theory of a democracy (of) thought and a communism (of) thought: "The democracy of-the-last-instance could after all be called 'communism'—if subtracted from every historical instance just as much as from spontanism, if the 'common' of communism was understood as the generic, if communism was understood as the generic constant of history" (François Laruelle, *Introduction aux sciences generiques* [Paris: Pétra, 2008] pp. 98–99).
5. Edmund Husserl, *The Crisis of European Sciences and Transcendental Phenomenology: An Introduction to Phenomenological Philosophy*, trans. David Carr (Evanston: Northwestern University Press, 1970), p. 189.
6. Readers interested in Laruelle's project may benefit from consulting *Laruelle and Non-Philosophy*, eds. John Mullarkey and Anthony Paul Smith (Edinburgh: Edinburgh University Press, 2012), as well as the works by Brassier and Mullarkey to be discussed.

236   NOTES

7. François Laruelle, *Philosophie et non-philosophie* (Mardaga: Liege-Bruxelles, 1989), p. 17.

8. See François Laruelle, "Theory of Philosophical Decision," in *Philosophies of Difference: A Critical Introduction to Non-Philosophy*, trans. Rocco Gangle (London and New York: Continuum, 2011), pp. 196–223; "Théorie de la Décision philosophique," in *Les Philosophies de la différence. Introduction Critique* (Paris: PUF, 1986), pp. 213–240; and "Analytic of Philosophical Decision," in *Principles of Non-Philosophy*, trans. Nicola Rubczak and Anthoy Paul Smith (London and New York: Bloomsbury, 2013), pp. 215–233; "Analytique de la Décision philosophique," in *Principes de la non-philosophie* (Paris: PUF, 1995), pp. 281–304.

9. Ray Brassier, *Nihil Unbound: Enlightenment and Extinction* (Basingstoke: Palgrave Macmillan, 2007), p. 123.

10. Ibid.

11. François Laruelle et al., *Dictionnaire de la non-philosophie* (Paris: Éditions Kimé, 1998), p. 40. See also Taylor Adkins's draft translation of this passage and the rest of the *Dictionnaire* available online: http://nsrnicek.googlepages .com/DictionaryNonPhilosophy.pdf. My own translation is modified from that of Adkins.

12. Brassier, *Nihil Unbound*, pp. 122–127.

13. Laruelle, *Dictionnaire de la non-philosophie*, p. 40.

14. Cf. Erik del Bufalo, *Deleuze et Laruelle. De la schizo-analyse à la non-philosophie* (Paris: Kimé, 2003), p. 34.

15. See Hugues Choplin, *De la phénoménologie à la non-philosophie. Lévinas et Laruelle* (Paris: Kimé, 1997); and Hugues Choplin, *L'espace de la pensée française contemporaine* (Paris: L'Harmattan, 2007); Eric Mollet, *Bourdieu et Laruelle. Sociologie réflexive et non-philosophie* (Paris: Éditions Petra, 2003); Didier Moulinier, *De la psychanalyse à la non-philosophie. Lacan et Laruelle* (Paris: Kimé, 1998); and Olivier Harlingue, *Sans condition. Blanchot, la littérature, la philosophie* (Paris: L'Harmattan, 2009); Patrick Fontaine, *Platon autrement dit* (Paris: L'Harmattan, 2007); Gilbert Kieffer, *Que peut la peinture pour l'esthétique?* (Paris: Éditions Petra, 2003); and *Esthétiques non-philosophiques* (Paris: Kimé, 1996); Anne-Françoise Schmid, "Le problème de Russell," in *La Non-philosophe des contemporains* (Paris: Kimé, 1995), pp. 167–186.

16. Ray Brassier, Iain Hamilton Grant, Graham Harman, and Quentin Meillassoux, "Speculative Realism," in *Collapse III* (Falmouth: Urbanomic, 2007), pp. 307–449. The transcript includes the four individual presentations by Brassier, Grant, Harman, and Meillassoux, and the discussion after each. Hereafter we will simply refer to this work as "Meillassoux, 'Speculative Realism.'"

17. Cf. Brassier, *Nihil Unbound*, p. 118.

18. Quentin Meillassoux, *After Finitude: An Essay on the Necessity of Contingency*, trans. Ray Brassier (New York and London: Continuum, 2008), p. 28.

19. I note in passing Brassier's characteristically harsh appraisal of Laruelle where he claims, "Laruelle's writings have yet to inspire anything beyond uncritical emulation or exasperated dismissal" (Brassier, *Nihil Unbound*, p. 118). Brassier

NOTES 237

places the reader of Laruelle in a gilded cage here perhaps to present his own reading of Laruelle as a critical use of non-philosophy, but this forecloses the third possibility of attempting to understand and use non-philosophy. To one who is hostile to Laruelle's non-philosophy this may already look like "uncritical emulation," and though Brassier himself is passing judgment here, he does not provide us with any metaphilosophy for judging what uncritical emulation looks like. It is important to keep in mind, as I will try to show later in the essay, that there is a difference here between method and content. One can "emulate" the method of non-philosophy and reject, even on non-philosophical grounds, the content Laruelle constructs much as Brassier himself does, albeit Brassier does so on standard philosophical grounds. My own positive non-philosophical project will, for instance, take a very different form from Laruelle's humanism (or, rather, non-humanism) on the basis of an understanding of nature that is neither purely Greek (cosmos, physis) nor purely Jew (creation) and that does not aim to bring them together into some kind of Christian Jew-Greek synthesis as World.

20. Meillassoux, "Speculative Realism," p. 427.
21. Meillassoux defines the thing-in-itself as the thing "independently of its relation to me" (Meillassoux, *After Finitude*, p. 1). He goes on from here to say "all those aspects of the object that can be formulated in mathematical terms can be meaningfully conceived as properties of the object in itself" (p. 3).
22. Meillassoux, "Speculative Realism," p. 419.
23. Cf. Laruelle, *En tant qu'Un*, p. 37.
24. Laruelle, *Philosophie et non-philosophie*, p. 16.
25. Cf. Laruelle, *Philosophies of Difference*, pp. 152–155; *Les philosophies de la différence*, pp. 169–172.
26. Bufalo, *Deleuze et Laruelle*, p. 17.
27. Meillassoux, "Speculative Realism," pp. 417–419. The argument appears popular and convincing among readers of Meillassoux, though it is far from clear that these are also readers of Laruelle based on their presentation of his work. Cf. Graham Harman, *Prince of Networks: Bruno Latour and Metaphysics* (Melbourne: re:press, 2009), pp. 177–178.
28. Brassier, *Nihil Unbound*, p. 128.
29. Laruelle, *Philosophie et non-philosophie*, pp. 176–177.
30. François Laruelle, *Théorie des identités. Fractalité généralisée et philosophie artificielle* (Paris: PUF, 1992), p. 59.

6 THE PRACTICE AND PRINCIPLES OF NON-PHILOSOPHY

1. François Laruelle, personal communication, May 2, 2010. Waves and phases, in the sense these words have within physics, become important concept in his most recent *Philosophie non-standard. Générique, quantique, philo-fiction*

(Paris: Kimé, 2010). Even though Laruelle aims not to separate the practice of non-philosophy from its material some of his books do develop the practice of non-philosophy with more clarity and attention. His most recent book is to be counted among these alongside of (in chronological order of original publication) *Philosophies of Difference: A Critical Introduction to Non-Philosophy*, trans. Rocco Gangle (London and New York: Continuum, 2010); *Philosophie et non-philosophie* (Mardaga: Liege-Bruxelles, 1989); *Théorie des identités*. *Fractalité généralisée et philosophie artificielle* (Paris: PUF, 1992); *Principles of Non-Philosophy*, trans. Nicola Rubczak and Anthony Paul Smith (London and New York: Bloomsbury, 2013); *Introduction au non-marxisme* (Paris: PUF, 2000); and *Future Christ: A Lesson in Heresy*, trans. Anthony Paul Smith (London and New York: Continuum, 2010).

2. Laruelle, *Principles of Non-Philosophy*, p. 33; *Principes de la non-philosophie* (Paris: PUF, 1995), pp. 38–39.

3. Laruelle, *Principles of Non-Philosophy*, p. 33; *Principes de la non-philosophie*, p. 39.

4. Laruelle, *Principles of Non-Philosophy*, p. 33; *Principes de la non-philosophie*, p. 39.

5. Laruelle, *Principles of Non-Philosophy*, p. 33; *Principes de la non-philosophie*, p. 39.

6. See the entry "Vision-en-Un (Un, Un-en-Un, Réel)" in Laruelle et al., *Dictionnaire de la non-philosophie* (Paris: Kimé, 1998), pp. 202–205.

7. Laruelle, *Philosophie et non-philosophie*, p. 38.

8. Laruelle, *Principles of Non-Philosophy*, p. 34; *Principes de la non-philosophie*, p. 39.

9. "In philosophy, Marxism included, immanence is an objective, a proclamation, an object, never a manner of thinking or a style" (Laruelle, *Introduction au non-marxisme*, p. 40).

10. See Erik del Bufalo, *Deleuze et Laruelle. De la schizo-analyse à la non-philosophie* (Paris: Kimé, 2003), p. 40.

11. Laruelle, *Philosophies of Difference*, pp. 198–202; *Les philosophies de la différence*, pp. 215–219. See also Laruelle, *Philosophies of Difference*, pp. 219–223; Laruelle, *Les philosophies de la différence*, pp. 237–240 for an early formal schema of the One. Cf. Laruelle, *Dictionnare*, pp. 202–205; and *Principles of Non-Philosophy*, 119–146; Laruelle, *Principes de la non-philosophie*, pp. 168–192.

12. Laruelle, *Principles of Non-Philosophy*, p. 34; *Principes de la non-philosophie*, p. 38.

13. Laruelle, *Principles of Non-Philosophy*, p. 34; *Principes de la non-philosophie*, p. 40.

14. Laruelle, *Principles of Non-Philosophy*, pp. 34–34; *Principes de la non-philosophie*, pp. 40–41. I read the parentheses framing the "of" to suggest that this is a unified relationship between force and thought rather than one being primary over the other. Thus the substantial meaning of the "of" is suspended. In my own creation of the population (of) thought and ecosystem (of) thought I make use of this parenthetical, recognizing as I do so that it can appear distracting and pretentious. I can only ask the reader's charity in reading it as a technical use of syntax that indicates this suspension.

15. See the entry "Force (de) pensée (sujet-existant-Étranger)" in Laruelle, *Dictionnaire*, pp. 76–79.

16. Ibid., p. 77.

17. Laruelle, *Principles of Non-Philosophy*, p. 35; *Principes de la non-philosophie*, p. 41.

NOTES   239

18. For other works in non-philosophy outside of Laruelle's development, see
    the bibliography of the other members of l'Organisation Non-philosophique
    Internationale. Available online: http://www.onphi.net/biblio/auteurs.php
    (accessed September 10, 2010).
19. François Laruelle, *Struggle and Utopia at the End Times of Philosophy*, trans.
    Drew S. Burk and Anthony Paul Smith (Minneapolis: Univocal Publishing,
    2012), p. 45; François Laruelle, *La Lutte et l'Utopie à la fin des temps philoso-
    phiques* (Paris: Éditions Kimé, 2004), p. 43.
20. Laruelle, *Philosophie non-standard*, p. 135.
21. Ibid., p. 137.
22. Ibid.
23. Alain Badiou, "Mathematics and Philosophy: The Grand Style and the
    Little Style," in *Theoretical Writings*, eds. and trans. Ray Brassier and Alberto
    Toscano (London and New York: Continuum, 2006), p. 17; emphases mine.
24. Iain Hamilton Grant, *Philosophies of Nature after Schelling* (London and New
    York: Continuum, 2006), p. 17.
25. Gilles Deleuze and Félix Guattari, *What is Philosophy?*, trans. Hugh
    Tomlinson and Graham Burchell (New York: Columbia University Press,
    1994), pp. 33, 37.
26. Ibid., p. 42.
27. Ibid., p. 161.
28. Laruelle, *Théorie des identités*, p. 56.
29. Ibid., p. 57.
30. Ibid., p. 58.
31. Ibid. I will explain the meaning of this "cause," which Laruelle calls
    "Identity-of-the-last-instance" later.
32. Ibid., p. 54.
33. Ibid.
34. Ibid., p. 55.
35. Ibid.
36. Ibid., p. 59.
37. Ibid., p. 60.
38. Ibid., p. 100.
39. Ibid.
40. François Laruelle, *Introduction au non-marxisme* (Paris: PUF, 2000), p. 62.
41. See Louis Althusser, *For Marx*, trans. Ben Brewster (London: Verso, 2005),
    pp. 104–106. When discussing the Marxist conception I have written out the
    phrase without the dashes, and when talking about the non-philosophical
    version I will include the dashes, as found in the original authors.
42. Ibid., p. 117.
43. See Ibid., pp 117–118, for a more technical discussion of these effects that put
    it into a quasi-Spinozist form.
44. Ibid., p. 99. Cf. Laruelle, *Dictionnare*, p. 48.
45. Laruelle, *Principles of Non-Philosophy*, p. 127; *Principes de la non-philosophie*,
    p. 152.
46. Laruelle, *Introduction au non-marxisme*, p. 40.

240 NOTES

47. John Mullarkey, *Post-Continental Philosophy: An Outline* (London and New York: Routledge, 2006), p. 149.
48. For Laruelle's full outline of this history, see his *Introduction aux sciences génériques* (Paris: Éditions Pétra, 2008), pp. 39–45.
49. Ibid., p. 44.
50. See ibid., pp. 41–43.
51. Ibid., p. 56.
52. Ibid., p. 57.
53. Ibid., p. 59.
54. Ibid.
55. François Laruelle, "From the First to the Second Non-Philosophy," in *From Decision to Heresy: Experiments in Non-Standard Thought*, ed. Robin Mackay, trans. Anthony Paul Smith and Nicola Rubczak (Falmouth: Urbanomic, 2013), p. 308.
56. Laruelle, *Introduction aux sciences génériques*, p. 14.
57. Laruelle, *Théorie des identités*, p. 56.
58. Laruelle, "From the First to the Second Non-Philosophy," p. 308.
59. See Laruelle, *Introduction aux sciences génériques*, p. 58.
60. Laruelle, "From the First to the Second Non-Philosophy," pp. 309, 308.

7 NON-THEOLOGICAL SUPPLEMENT

1. Alain Badiou, interview with Ben Woodward, *The Speculative Turn: Continental Materialism and Realism*, eds. Levi Bryant, Nick Srnicek, and Graham Harman (Melbourne: re.press, 2011), p. 20.
2. Of course Laruelle has made his own remarks about Badiou. See François Laruelle, *Anti-Badiou: On the Introduction of Maosim into Philosophy*, trans. Robin Mackay (London and New York: Bloomsbury, 2013).
3. François Laruelle, *Future Christ: A Lesson in Heresy*, trans. Anthony Paul Smith (London and New York: Continuum, 2010), p. 15; *Le Christ futur. Une leçon d'hérésie* (Paris: Exils Éditeur, 2002), p. 31.
4. François Laruelle, *Mystique non-philosophique à l'usage des contemporains* (Paris: L'Harmattan, 2007), p. 33.
5. Laruelle, *Future Christ*, p. 27; *Le Christ futur*, p. 44.
6. See Laruelle, *Future Christ*, p. 29; *Le Christ futur*, p. 47.
7. Ibid.
8. Jean-Luc Rannou, *La non-philosophie, simplement. Une introduction synthétique* (Paris: L'Harmattan, 2005), p. 114.
9. It goes without saying that his democracy is real and transcendent to any kind or form of representational democracy that remains subsumed in politics rather than enacting any kind of real democracy. Much the same as the communism Laruelle embraced in a previous chapter is real and transcendent to any kind of state-form of communism. See ibid., p. 76.

NOTES   241

10. François Laruelle, "A New Presentation of Non-Philosophy," working paper, L'Organisation Non-Philosophique Internationale, February 11, 2004. Available online: http://www.onphi.net/texte-a-new-presentation-of-non-philosophy-32.html (accessed February 1, 2010).
11. Ibid.
12. Janicaud's criticism of the theological turn as beginning from a method of "take it or leave it" suggests an axiomatic character to the theological turn as well and so suggests that Laruelle is infected by the same philosophical illness as Henry and Marion. Dominique Janicaud, *Phenomenology "Wide Open": After the French Debate*, trans. Charles N. Cabral (New York: Fordham University Press, 2005), pp. 4–7.
13. This is one of the most refreshing aspects of Laruelle's passage from Philosophy I to V, the upfront recognition of his own works' failures and inadequacies as non-philosophy is continually performed anew in the light of new material. See again François Laruelle, *Principes de la non-philosophie* (Paris: PUF, 1995), pp. 19–42; and *Philosophie non-standard. Générique, quantique, philo-fiction* (Paris: Kimé, 2010), pp. 15–44.
14. See Philip Goodchild, *Capitalism and Religion: The Price of Piety* (London: Routledge, 2002), pp. 51–57. Here Goodchild traces the ways that the Christian Creeds were codified into a "metaphysical, universal, eternal spiritual and written" truth that could then be exchanged across cultures (p. 53).
15. See Laruelle, *Future Christ*, pp. 140–143; *Le Christ futur*, pp. 173–176.
16. Laruelle, *Philosophie et non-philosophie* (Mardaga: Liege-Bruxelles, 1989), pp. 176–177.
17. John Mullarkey, *Post-Continental Philosophy: An Outline* (London and New York: Routledge, 2006), p. 140.
18. See François Laruelle, *Une Biographie de l'homme ordinaire. Des Autorités et des Minorités* (Paris: Aubier, 1985), pp. 7–38. There he remarks that "[m]an has never been the object of the human Sciences" (p. 8). Rather Man is subsumed into some other aspect of the World in the search for a logic of man (the anthropo-logical is the amphibology of logic and humanity). It should also be noted that it is difficult to express in a single word the meaning of *individual* in English as Laruelle differentiates *individual* from *individuel*. The first is a neologism of his own construction playing on the sense of "dual" in order to express the fundamental duality of individuals, and the second is the term that is usually translated into English as individual, but he plays with the "duel," which means the same in English as it does in French, to signify that it is a fundamentally antagonistic concept. See ibid., p. 17.
19. Laruelle, *Future Christ*, p. 34; *Le Christ futur*, p. 52.
20. Laruelle, *Future Christ*, pp. 42–43; *Le Christ futur*, pp. 60–61.
21. Laruelle, *Future Christ*, p. 4; *Le Christ futur*, p. 22.
22. Laruelle, *Future Christ*, p. 7; *Le Christ futur*, p. 25.
23. Laruelle, *Future Christ*, pp. 4, 8; *Le Christ futur*, pp. 22, 26.
24. See Henry Corbin, *Le Paradoxe du Monothéisme* (Paris: L'Herne, 2003) for a concise and convincing study of this dual character of monotheistic religions.

242   NOTES

25. François Laruelle, *Struggle and Utopia at the End Times of Philosophy*, trans. Drew S. Burk and Anthony Paul Smith (Minneapolis: Univocal Publishing, 2012), p. 250; *La Lutte et l'Utopie à la fin des temps philosophiques* (Paris: Kimé, 2004), p. 204.
26. John Milbank, "Knowledge: The Theological Critique of Philosophy in Hamann and Jacobi," in *Radical Orthodoxy: A New Theology*, eds. John Milbank, Catherine Pickstock, and Graham Ward (London: Routledge, 1999), p. 21.
27. Talal Asad, *Formations of the Secular: Christianity, Islam, Modernity* (Stanford: Stanford University Press, 2003), p. 5.
28. Laruelle, *Philosophie non-standard*, p. 125.
29. Laruelle, *Future Christ*, p. 117; *Le Christ futur*, p. 145.
30. Zachery Luke Fraser, Draft of "Entry for 'Generic,'" in *The Badiou Dictionary*, ed. Steven Corcoran (Edinburgh: Edinburgh University Press, forthcoming). Draft available online: http://formandformalism.blogspot.com/2011/03/generic-entry.html (accessed March 25, 2011).

8 REAL ECOSYSTEMS (OF) THOUGHT

1. On this, see Daniel Colucciello Barber, *Deleuze and the Naming of God: Post-Secularism and the Future of Immanence* (Edinburgh: Edinburgh University Press, forthcoming) and my own essay, which builds upon Barber's work to describe Deleuze and Guattari's ecological philosophy of nature, "Believing in this World for the Making of Gods: On the Ecology of the Virtual and the Actual," *SubStance* 38.3 (April 2010): 101–112.
2. Christian Lévêque, *Ecology: From Ecosystem to Biosphere* (Plymouth, UK: Science Publisher, Inc., 2003), p. 29.
3. Brian Walker and David Salt, *Resilience Thinking: Sustaining Ecosystems and People in a Changing World* (Washington and London: Island Press, 2006), p. xiii.
4. François Laruelle, *Théorie des identités* (Paris: PUF, 1992), p. 56.
5. François Laruelle, *Philosophie non-standard. Générique, Quantique, Philo-fiction* (Paris: Kimé, 2010), p. 54.
6. See Anthony Paul Smith, "Philosophy and Ecosystem: Towards a Transcendental Ecology," *Polygraph* 22 (2010): 65–82.
7. Rocco Gangle, "Translator's Introduction," in François Laruelle's *Philosophies of Difference: A Critical Introduction to Non-Philosophy*, trans. Rocco Gangle (London and New York: Continuum, 2010), p. vi.
8. Laruelle, *Philosophie non-standard*, p. 490.
9. Ibid.
10. Lévêque, *Ecology*, p. 8.
11. Cf. ibid., p. 9. Which isn't to say that management falls outside of ecology as such, just that the management itself should be subordinated to knowledge rather than the demands of quick and popular policies.

NOTES   243

12. Slavoj Žižek, "Censorship Today: Violence, or Ecology as a New Opium
    for the Masses." Available online: http://www.lacan.com/zizecology1.htm
    (accessed September 10, 2010).
13. François Laruelle, "L'Impossible foundation d'une écologie de l'océan." Available
    online: http://www.onphi.net/lettre-laruelle-l-impossible-fondation-d-une
    -ecologie-de-l-ocean-27.html (accessed March 22, 2011).
14. Ibid.
15. Ibid. Laruelle often indicates he is talking about a false transcendent version
    of a concept by emphasizing the definite article. This works better in French
    than in English, as "the nature" is not idiomatic. Often, though, I am forced
    to translate it this way to retain Laruelle's meaning. In this case, however, a
    capital "N" serves the same purpose.
16. Ibid. For a historical discussion of the difficulties capitalism has encoun-
    tered with the ocean as regards property rights, see Joachim Radkau, *Nature
    and Power: A Global History of the Environment*, trans. Thomas Dunlap
    (Cambridge: Cambridge University Press, 2008), pp. 86–93.
17. Laruelle, "L'Impossible foundation d'une écologie de l'océan."
18. Lévêque, *Ecology*, p. 8.
19. Laruelle, "L'Impossible foundation d'une écologie de l'océan."
20. François Laruelle, "The Degrowth of Philosophy: Toward a Generic Ecology,"
    trans. Robin Mackay in *From Decision to Heresy: Experiments in Non-Standard
    Thought*, ed. Robin Mackay (Falmouth: Urbanomic, 2013) p. 328.
21. Ibid., p. 337–338.
22. Ibid., p. 340.
23. Ibid., p. 345.
24. Ibid., p. 346.
25. Ibid.
26. Ibid., p. 349.

9   ELEMENTS OF AN IMMANENTAL ECOLOGY

1. François Laruelle, *Philosophie non-standard. Générique, quantique, philo-fiction*
   (Paris: Kimé, 2010), pp. 53–54.
2. Timohy Morton, *The Ecological Thought*, (London and Cambridge, MA:
   Harvard University Press, 2010), p. 3.
3. See Ibn Khaldûn, *The Muqaddimah: An Introduction to History*, trans. Franz
   Rosenthal (Princeton: Princeton University Press, 1969), pp. 49–64; Jared
   Diamond, *Guns, Germs and Steel: A Short History for the Last 13,000 Years*
   (London: Vintage, 2005); and *Collapse: How Societies Choose to Succeed or
   Fail* (London and New York: Penguin, 2006); Donald Worster, *Nature's
   Economy: A History of Ecological Ideas*, 2nd edition (Cambridge: Cambridge
   University Press, 1994); Joachim Radkau, *Nature and Power: A Global
   History of the Environment*, trans. Thomas Dunlap (Cambridge: Cambridge

244   NOTES

University Press, 2008); Alfred W. Crosby, *Ecological Imperialism: The Biological Expansion of Europe, 900–1900* (Cambridge: Cambridge University Press, 1986).

4. Cf. Worster, *Nature's Economy*, Part IV.

5. Lévêque summarizes this debate nicely writing, "Ecology is especially the field of heuristic principles, that is, hypotheses that we do not seek to prove as true or false, but that are adopted provisionally as directive ideas in search for facts. Unlike physics or genetics, for example, ecology has not created a significant construct of organized laws. In this sense, many propositions of ecology, such as the concepts of niche, climax, or even biosphere, cannot be tested in the sense that Popper defines testing. But one of the reasons why we sometimes question the status of ecology as science is that it is often difficult to eliminate the particular point of view of the observer and to eliminate all value judgements on the object of study. In reality, behind this rather academic debate lies the major question of recognition of scientific domains that do not raise the same paradigms as those of physics, which have dominated science till recently. There are domains in which the elaboration of universal and deterministic laws, and the experimental process, are much more difficult to implement, taking into account their nature and complexity. Some even cast doubt on the fundamentals of scientific discourse and search for alternatives to the basic paradigms that have been offered to us by physics and mathematics. Behind the well-known affirmation that the whole is more than the sum of its parts lies the idea of an ecosystem that is not simply a juxtaposition of living and inert elements but has emerging properties that are not deduced from just the characteristics of its components. There lies the difficulty" (Christian Lévêque, *Ecology: From Ecosystem to Biosphere* [Plymouth, UK: Science Publisher, Inc., 2003], p. 4).

6. Golley explains in his book that population ecology was dominant in Britain, while ecosystem ecology is often said to have developed largely in America. He writes, "The study of species populations and the interaction between populations captured the attention of ecologists and became a major area of British ecological work. It was not only exciting in itself, but it satisfied the scientific desire for ecology to move to a phase where hypotheses were tested through experiment and observation. One could undertake experiments with populations, and it was possible to apply the hypothetical-deductive approach to them. Further, species population ecology built upon the long British history of fieldwork in natural history in which botanists or zoologists collected, described, named, and reported on the distribution and abundance of organisms. Large research teams were not required. The research could be carried out by an individual ecologist" (Frank Golley, *A History of the Ecosystem Concept in Ecology: More Than the Sum of the Parts* [New Haven and London: Yale University Press, 1993], p. 177).

7. Lévêque, *Ecology*, p. 5.

8. F. Stuart Chapin III, Pamela A. Matson, and Harold A. Mooney, *Principles of Terrestrial Ecosystem Ecology* (London: Springer, 2002), p. 7.

NOTES   245

9. Golley, *A History of the Ecosystem*, p. 1.
10. Ibid., p. 66.
11. Ibid.
12. Ibid., p. 69.
13. Ibid., pp. 71, 72.
14. Ibid., p. 36.
15. Ibid., pp. 37–39.
16. Ibid., p. 56.
17. Cf. François Laruelle, *Introduction aux sciences génériques* (Paris: Pétra, 2008), pp. 24–31.
18. Lévêque, *Ecology*, p. 282.
19. Edward O. Wilson, *The Diversity of Life* (London and New York: Penguin Books, 2001), pp. xii, xx.
20. Ibid., p. 36.
21. Ibid., p. x. Wilson goes on to discuss how even this unit throws up difficulties in terms of counting. For instance, he writes that "[o]ne third of all recognized species of organisms are parasites" and "[b]acteria continue to be the 'black hole' of biodiversity, their depths unplumbed" (p. xiii).
22. Ibid., p. 42.
23. Ibid.
24. Ibid, pp. 42–45. It should be noted that the majority of hybridization between species has thus far been found within plant species in the tropics.
25. This is one of the truly interesting aspects of Mullarkey's reading of Laruelle, as he reads Laruelle as fundamentally thinking from the actual rather than the virtual. See John Mullarkey, *Post-Continental Philosophy: An Outline* (London and New York: Continuum, 2006), pp. 125–137.
26. Wilson, *The Diversity of Life*, p. 46.
27. Gilles Deleuze, *Bergsonism*, trans. Hugh Tomlinson and Barbara Habberjam (New York: Zone Books, 1991), p. 29.
28. Paul S. Giller, *Community Structure and the Niche* (London and New York: Chapman and Hall, 1984), pp. 1, 9.
29. Ibid., p. 1.
30. Ibid., p. 7.
31. Cf. Wilson, *The Diversity of Life*, p. 217.
32. Ibid., pp. 196–198. Cf. Giller, *Community Structure and the Niche*, p. 10, where he writes "Each environmental gradient can be thought of as a dimension in space. If there are *n* pertinent dimensions the niche can be described in terms of an *n*-dimensional space, or hypervolume."
33. Giller, *Community Structure and the Niche*, p. 9.
34. Wilson, *The Diversity of Life*, p. 111.
35. Ibid., p. 217. Giller explains that the niche width is formed by the combination of two separate components—the within-phenotype component (WPC), which describes the level of variation in resource use by individuals, and the between-phenotype component (BPC), which describes the variation among individuals of the species population (p. 12). Broad niches mean that the species tends to use resources in proportion to their availability,

246 NOTES

while narrow niches mean that the species tends to concentrate on only some resources (p. 13).

36. Pierre Hadot, *The Veil of Isis: An Essay on the History of the Idea of Nature*, trans. Michael Chase (Cambridge, MA, and London: The Belknap Press of Harvard University Press, 2006), p. 10.

37. Antonio Negri, *The Labor of Job: The Biblical Text as a Parable for Human Labor*, trans. Matteo Mandarini (Durham and London: Duke University Press, 2009), p. 27. As is common with translations of Negri and Deleuze and Guattari, when power is spelled with a lowercase p it is translating the French *puissance* or the Italian *potenza* and when it is spelled with an uppercase P it translates the French *pouvoir* or the Italian *potere*.

38. Ibid., p. 28.

39. Ibid., p. 59.

40. Ibid., p. 75.

41. Ibid., p. 81.

42. Ibid., pp. 96–97.

43. Ibid., p. 52.

44. Cf. Ray Brassier, *Nihil Unbound: Enlightenment and Extinction* (Basingstoke: Palgrave Macmillan, 2007), p. 25.

45. cf. Negri, *Labor of Job*, pp. 50, 73, for a discussion of this idea as it is found in the Book of Job.

46. Cf. Giorgio Agamben, *The Open: Man and Animal*, trans. Kevin Attell (Stanford: Stanford University Press, 2004). Agamben's short text provides a succinct summary of the biopolitical elements at work in philosophy of nature. It should be noted that the emphasis on Jacob von Uexküll, found in Heidegger, Merleau-Ponty, and to a lesser extent Deleuze, as well as their interpreters, is out of proportion to the actual legacy he has left on scientific ecology. His name, e.g., is not mentioned in any of the major histories I have read for this project and yet Agamben mistakenly credits him as one of the "founders of ecology" (p. 39).

47. Chapin, Matson, and Mooney, *Principles of Terrestrial Ecosystem Ecology*, p. 4.

48. Ibid., p. 140.

49. Ibid., pp. 97, 140.

50. Golley, *A History of the Ecosystem*, p. 81.

51. Ibid., pp. 67–69.

52. Howard Thomas Odum quoted in ibid., p. 81.

53. Lévêque, *Ecology*, p. 246.

54. Chapin, Matson, and Mooney, *Principles of Terrestrial Ecosystem Ecology*, p. 10.

55. Ibid.

56. Lévêque, *Ecology*, p. 247.

57. Chapin, Matson, and Mooney, *Principles of Terrestrial Ecosystem Ecology*, p. 8.

58. Lévêque, *Ecology*, pp. 247, 249.

59. Brassier, *Nihil Unbound*, p. xi.

60. Chapin, Matson, and Mooney, *Principles of Terrestrial Ecosystem Ecology*, p. 10.

61. Ibid., p. 31.

NOTES   247

62. Ibid., p. 11.
63. Lévêque, *Ecology*, p. 265.
64. Ibid., p. 174.
65. Ibid. Eotones are an example of spatial heterogeneity.
66. Ibid.
67. Ibid.
68. Cf. Crosby, *Ecological Imperialism*.
69. Eric Katz, "The Big Lie: Human Restoration of Nature," *Research in Philosophy and Technology* 12 (1992), 232.
70. Ibid., p. 233.
71. See Ibid., p. 234.
72. Brian Walker and David Salt, *Resilience Thinking: Sustaining Ecosystems and People in a Changing World* (Washington and London: Island Press, 2006), p. 11.
73. Ibid., p. 9.
74. Ibid., p. 7–9.
75. Ibid., p. 9. The most revealing example of such management is to be found in the Goulburn-Broken Catchment in Southwest Australia. See pp. 39–52.
76. Ibid.
77. Ibid., pp. 9–10.
78. Ibid., p. 141.
79. Ibid., p. 11.
80. Ibid., pp. 53–55.
81. Ibid., p. 63.
82. Ibid.
83. Ibid., p. 75. One must keep in mind of course that the word "cycle" is a useful metaphor in describing change and not a given or a metaphysical claim.
84. Ibid., p. 76.
85. Ibid.
86. Ibid., p. 77.
87. Ibid., p. 78.
88. Ibid.
89. Ibid.

10   ECOLOGIES WITHOUT NATURE

1. Bruno Latour, *Politics of Nature: How to Bring the Sciences into Democracy*, trans. Catherine Porter (Cambridge, MA, and London: Harvard University Press, 2004), p. 245.
2. Ibid., p. 249.
3. Ibid., p. 248.
4. Ibid., p. 54.

248   NOTES

5. Cf. Ibid., p. 224.
6. Ibid., p. 37.
7. Ibid., pp. 10–18.
8. Ibid., p. 231.
9. See Joachim Radkau, *Nature and Power: A Global History of the Environment*, trans. Thomas Dunlap (Cambridge: Cambridge University Press, 2008), pp. 36–77.
10. Timothy Morton, *The Ecological Thought* (Cambridge, MA, and London: Harvard University Press, 2010), p. 104.
11. Timothy Morton, *Ecology without Nature: Rethinking Environmental Aesthetics* (Cambridge, MA, and London: Harvard University Press, 2007), pp. 92–94.
12. See Christian Jambet and Guy Lardreau, *L'Ange: Pour une cyégétique du semblant, Ontologie de la révolution 1* (Paris: Bernard Grasset, 1976), pp. 17–37.
13. Cf. Morton, *The Ecological Thought*, pp. 2–4.
14. Ibid., p. 2. Morton's specific engagements with ecology are not of particular importance here as my own engagement with scientific ecology is complementary to his notions (like that of "mesh," which is a slightly more literary way of saying "ecosystem"), though I think the engagement here is more sustained. Because of this I have elided any in-depth discussion of them, but they are succinctly summarized in *The Ecological Thought*.
15. Morton, *Ecology without Nature*, p. 7.
16. Ibid., p. 5.
17. Ibid., pp. 6, 52. Although Morton's presentation of Deleuze and Guattari's "rhizome" seems to have a weak understanding of the diversity of different rhizomatic plants, such as European grasses and "weeds," and his suggestion that they don't have to deal with gravity is confused, his criticisms regarding the collapse of difference into identity suggest there is a parallel between Laruelle's critique of difference and his own.
18. Morton, *The Ecological Thought*, p. 3.
19. Morton, *Ecology without Nature*, p. 21.
20. Radkau's argues for the same idea, showing how the conception of nature operative in modern environmentalism is parallel to the Christian idea of an omnipotent God (Radkau, *Nature and Power*, p. 83).
21. Morton, *Ecology without Nature*, p. 21.
22. Morton, *The Ecological Thought*, p. 98. Cf. pp. 98–101.
23. Ibid., p. 3.
24. In the American West the intentional killing of animals considered to be "varmints" resulted in a number of ecological issues. These included the usual maligned species, rodents said to carry disease or destroy human crops, and also larger mammals such as wolves (who have been all but wiped out in the American West) and coyotes. See Donald Worster, *Nature's Economy: A History of Ecological Ideas*, 2nd edition (Cambridge: Cambridge University Press, 1994), Chapter 13.
25. Laruelle, *Philosophie non-standard*, pp. 80–81. The bracketed "the" before science appears to be Laruelle's attempt to indicate that the idealist form of science is here being bracketed.

NOTES   249

11 SEPARATING NATURE FROM THE WORLD

1. Consider the typical way it is used by Christian theologians in one-sided arguments (since they seem to be the only ones thinking an argument is going on)
with secular philosophers. One of the most direct examples is the writing of
American Methodist Daniel M. Bell Jr.: "Deleuze embraced the univocity
of being in the name of nurturing difference and freedom. Yet as Aquinas
lays out the analogy and the univocity of being in the *Summa Theologica*, it
becomes clear that the univocity of being cannot preserve difference—at least
it cannot preserve difference peaceably, which means that it cannot preserve
freedom. [...] The only mediating position for univocal being is conflict,
competition, combat. Univocal being can maintain difference in relation only
by means of a friction (agony) created between different degrees of intensity that necessarily mediates the encounter and clash of otherwise univocal
beings. In contract, the analogy of being maintains difference in a way that
allows difference to be drawn into a relation while preserving (and in the
case of the human, enhancing) the freedom of being. Only the analogy of
being permits differences to draw near in a mode other than competition and
conflict such that in this embrace of intimacy neither being nor its properties are lost (recall Chalcedon, or the Thomistic principle that grace does not
destroy but perfects nature)" (Daniel M. Bell, Jr., "Only Jesus Saves: Toward a
Theo-political Ontology of Judgment," in *Theology and the Political: The New
Debate*, eds. Creston Davis, John Milbank, and Slavoj Žižek (Durham: Duke
University Press, 2005), p. 216).
2. See Mark D. Jordan, *Rewritten Theology: Aquinas after His Readers* (Oxford:
Blackwell, 2006), pp. 1–17.
3. Regarding the relationship of truth and politics Jordan summarizes Aquinas's
position, "no ruler can prescribe doctrine. The city may regulate the circum-
stances of teaching, never its truth" (ibid., p. 179).
4. Philip Goodchild, *Capitalism and Religion: The Price of Piety* (London and
New York: Routledge, 2002), p. 210.
5. For an excellent intellectual biography, see Jean-Pierre Torrell O. P., *Saint
Thomas Aquinas*, trans. Robert Royal, vol. I: *The Person and His Work*
(Washington, D.C.: Catholic University of America Press, 1996). For an
equally excellent account of Aquinas's spiritual life, see Jean-Pierre Torrell
O. P., *Saint Thomas Aquinas*, trans. Robert Royal, vol. II: *Spiritual Master*
(Washington D.C.: Catholic University of America Press, 2003).
6. See Torrell, *Saint Thomas Aquinas*, vol. I, pp. 1–17.
7. Ibid., p. 73.
8. Ibid., pp. 36–53, 75–98.
9. See Bernard Montagnes, *The Doctrine of the Analogy of Being According to
Thomas Aquinas*, trans. E. M. Macierowski (Milwaukee, WI: Marquette
University Press, 2004), pp. 9, 12. This former function of the analogy is
what Montagnes calls the predicamental (how we speak of beings) while the
latter is the transcendental analogy or the account of being.

250 NOTES

10. Ibid., p. 66.
11. For an interesting and in-depth history of the early reception of Spinoza's works, see Jonathan I. Israel, *Radical Enlightenment: Philosophy and the Making of Modernity 1650–1750* (Oxford: Oxford University Press, 2001), Chapter 16.
12. Benedict de Spinoza, *Theological-Political Treatise*, trans. Michael Silverthorne and Jonathan Israel (Cambridge: Cambridge University Press, 2007), p. 15.
13. Goodchild, *Capitalism and Religion*, p. 75.
14. Ibid., p. 73. What Goodchild means by cynical is important for our understanding of Spinoza's philosophy of nature. The conception of cynicism that Goodchild is using here comes from his reading of Diogenes the Cynic. The essence of cynicism is there shown in the recounting of two stories; the first is Diogenes responding to Plato's definition of man as a featherless biped by bringing in a plucked fowl into the lecture room and the second is Diogenes drawing of attention away from a set speech by eating lupins followed by his feigned shock that the assembly would be distracted from the speech by the simple eating of lupins. "Such cynical gestures consist in responding to questions concerning the highest ideals with something material, edible and mortal" (p. 71).
15. Christian Lévêque, *Ecology: From Ecosystem to Biosphere* (Plymouth, UK: Science Publisher, Inc., 2003), p. 353.
16. Ibid., p. 357.
17. Anne Primavesi, *Sacred Gaia: Holistic Theology and Earth System Science* (London and New York: Routledge, 2000), p. xii.
18. Ibid.
19. Karl Marx, "Theses on Feuerbach," in *The Marx and Engels Reader*, ed. Robert C. Tucker (New York and London: WW Norton & Company, 1978), p. 145.
20. Martin Heidegger, "Modern Science, Metaphysics, and Mathematics," in *Basic Writings*, ed. David Farrell Krell (New York: HarperSanFrancisco, 1993), p. 276; and "…Poetically Man Dwells…," in *Poetry, Language, Thought*, trans. Albert Hofstadter (New York: Harper Perennial Modern Classics, 1971), p. 213.
21. Heidegger, "Modern Science," pp. 291, 273.
22. Ibid., p. 303.
23. Martin Heidegger, *The Fundamental Concepts of Metaphysics: World, Finitude, Solitude*, trans. William McNeill and Nicholas Walker (Bloomington: Indiana University Press, 1995), p. 5.
24. Ibid.
25. Ibid.
26. Ibid.; emphasis mine.
27. See ibid., p. 25.
28. Ibid.
29. See Brett Buchanan, *Onto-Ethologies: The Animal Environments of Uexküll, Heidegger, Merleau-Ponty, and Deleuze* (Albany: SUNY Press, 2008).
30. Martin Heidegger, *Being and Time*, trans. John Macquarrie and Edward Robinson (New York: HarperSanFrancisco, 1962), p. 78.
31. Ibid.

NOTES   251

32. Ibid., pp. 79–80.
33. Buchanan, *Onto-Ethologies*, p. 54. Cf. Graham Harman, *The Quadruple Object* (Winchester, UK: Zero Books, 2011), p. 86.
34. Heidegger, *Fundamental Concepts*, p. 184.
35. Ibid., p. 29.
36. Ibid., p. 28.
37. Heidegger, "...Poetically Man Dwells...," p. 212.
38. Ibid., p. 213.
39. Alain Badiou, *Being and Event*, trans. Oliver Feltham (New York and London: Continuum, 2005), p. 4.
40. Heidegger, *Fundamental Concepts*, p. 22. See pp. 1–23 for his prolonged discussion of the differentiation of philosophy from science and worldview as well as its essence as rooted in the experience of Dasein.
41. Alain Badiou, *Manifeste pour la philosophie* (Paris: Éditions du Seuil, 1980), p. 15. He repeats these conditions from *Being and Event*, but where in *Being and Event* he has "art" this changes in *Manifeste pour la philosophie* to simply "poetry." See Badiou, *Being and Event*, p. 4.
42. Badiou, *Manifeste*, p. 19.
43. See ibid., p. 23, where he discusses the errors of different philosophers in elevating different truth-procedures to the essence of philosophy, suturing philosophy to that truth-procedure in a way that leads to errors, including the error, in Badiou's eyes, of understanding ontology via poetry.
44. Badiou, *Being and Event*, p. 4.
45. See ibid., pp. 123–129.
46. Ibid., p. 125.
47. Ibid., pp. 125–126.
48. Ibid., p. 126.
49. Ibid., p. 127.
50. Badiou does not present his system in this way, the question is never on the particular actuality but always on the truth-event that recreates that thing. See Alain Badiou, *Logics of Worlds: Being and Event II*, trans. Alberto Toscano (New York and London: Continuum, 2009), pp. 9–33. Here Badiou sketches out how, in his view, the particular truth that "[t]here are only bodies and language; except that there are truths" manifests in the four conditions of philosophy.
51. Ibid., p. 127.
52. Ibid., p. 129.
53. Ibid., pp. 132–134.
54. Ibid., p. 140.
55. Badiou, *Logics of Worlds*, p. 4.
56. Ibid.
57. Badiou, *Being and Event*, p. 127.
58. Badiou, *Logics of Worlds*, p. 9.
59. Ibid., p. 598.
60. Cf. Michael Lewis, *Heidegger beyond Deconstruction: On Nature* (London and New York: Continuum, 2007). Lewis shows that, despite the important

252   NOTES

differences between the early and later work of Heidegger, the concept of
World and Being are continually at play.

61. Martin Heidegger, "The Thing," in *Poetry, Language, Thought*, trans. Albert
    Hofstadter (New York: Harper Perennial Modern Classics, 1971), p. 178.
62. Ibid., p. 175.
63. Cf. Lewis, *Heidegger beyond Deconstruction*, p. 10.
64. Harman, *The Quadruple Object*, pp. 86–87.
65. Cf. ibid., pp. 96–99.
66. Julian Young, "The Fourfold," in *The Cambridge Companion to Heidegger*, ed.
    Charles B. Guignon, 2nd edition (Cambridge: Cambridge University Press,
    2006), p. 375.
67. Lewis, *Heidegger beyond Deconstruction*, p. 10.
68. Young, "The Fourfold," p. 374. Cf. Heidegger, "...Poetically Man Dwells...,"
    pp. 149, 178.
69. Heidegger, "...Poetically Man Dwells...," p. 148.
70. Ibid., p. 182.
71. Ibid., p. 220.
72. Now, bringing to mind Laruelle's focus on the ocean discussed in chapter 8,
    here the term "earth" refers really to the biosphere, which includes all the
    wave-like aspects of energy flowing in ecosystems through the materials that
    make them up.

12   MATERIALS FOR A THEORY OF NATURE

 1. Rudi te Velde, *Aquinas on God: The "Divine Science" of the Summa theologiae*
    (Aldershot: Ashgate, 2006), p. 85.
 2. St. Thomas Aquinas, *Summa Theologica* (literal English translation), trans.
    Fathers of the English Dominican Province (Westminster, MD: Christian
    Classics, 1981), Ia, q. 4, a. 2.
 3. Velde, *Aquinas on God*, p. 85.
 4. St. Thomas Aquinas, *Summa Contra Gentiles*, trans. A. C. Pegis, James F.
    Anderson, Vernon J. Bourke, and Charles J. O'Neil, 4 vols. (Notre Dame, IN:
    University of Notre Dame Press, 1975), I/30; hereafter SCG.
 5. SCG I/30.3.
 6. Ibid.
 7. SCG I/30.2.
 8. SCG I/30.4. In this passage Aquinas only lists "infinite." Cf. Velde, *Aquinas
    on God*, p. 80, for a discussion of the other three in the *Summa Theologica*.
 9. Ibid.
10. Ibid.
11. SCG I/32.4. The argument against univocity is philosophically weak as it is
    worked out from a position where the metaphysics of God have already been
    decided theologically through the demands of the creeds and the Church
    Fathers.

NOTES   253

12. SCG I/32.5.
13. SCG I/33.2.
14. SCG I/33.3.
15. SCG I/33.5.
16. See SCG I/34.2–4. Here Aquinas recognizes that analogy could fall into a similar trap as the other grammars where there is a third term that one must recognize prior to both God and creation. He himself employs a variation of this form of analogy in his *De principiis naturae*, which we explore later.
17. SCG I/34.6.
18. See Anthony Kenny, *Aquinas on Being* (Oxford: Oxford University Press, 2002).
19. St. Thomas Aquinas, "Commentary on Boethius' *De Trinitate* [selection]," in *Selected Philosophical Writings*, ed. and trans. Timothy McDermott (Oxford: Oxford University Press, 1993), p. 44.
20. Ibid.
21. Ibid.
22. Ibid., p. 45.
23. Velde, *Aquinas on God*, p. 118.
24. See Bernard Montagnes, *The Doctrine of the Analogy of Being According to Thomas Aquinas*, translated by E. M. Macierowski (Milwaukee, WI: Marquette University Press, 2004), p. 24.
25. St. Thomas Aquinas, "On the Principles of Nature," in *Selected Philosophical Writings*, ed. and trans. Timothy McDermott (Oxford: Oxford University Press, 1993), p. 67.
26. Aquinas's Latin *natura* has more to do with its relation to "birth" or the generation of things than it does with the cosmos or universe, which would be indicated under the name *mundus*. It would appear that, for Aquinas, *natura* names the way in which the *mundus* persists in being. The contemporary slippage between World and Nature is thus already present in the Medieval sources.
27. Simply put matter names whatever exists potentially (Aquinas, "On the Principles of Nature", p. 67). Form names the actualization of the thing from its potential as matter (p. 68). Lack of form is an incidental principle, which is still necessary to generation as matter must lack form to begin with (p. 69).
28. Ibid., pp. 70–71.
29. Ibid., p. 71.
30. Ibid.
31. Aquinas, "Commentary on Botheius' *De Trinitate*," p. 45.
32. See John Milbank and Catherine Pickstock, "Truth and Correspondence," in *Truth in Aquinas* (London: Routledge, 2001).
33. SCG I/65.2
34. Milbank and Pickstock, "Truth and Correspondence," p. 15.
35. St. Thomas Aquinas, "On Kingship or The Governance of Rulers (De Regimine Principum, 1265–1267)," in *St. Thomas Aquinas on Politics and Ethics*, ed. and trans. Paul E. Sigmund (London and New York: W. W. Norton & Company, 1988), pp. 17–18.

254 NOTES

36. François Zourabichvili, *Spinoza. Une physique de la pensée* (Paris: PUF, 2002), p. 255.
37. Ibid., p. 220.
38. Rocco Gangle, "Theology of the Chimera: Spinoza, Immanence, Practice," in *After the Postsecular and the Postmodern: New Essays in Continental Philosophy of Religion*, eds. Anthony Paul Smith and Daniel Whistler (Newcastle-upon-Tyne: Cambridge Scholars' Press, 2010), p. 27.
39. Ibid.
40. For example, Steven Nadler's biography of Spinoza, *Spinoza: A Life* (Cambridge: Cambridge University Press, 1999), which presents a radically atheist Spinoza, as does Stephen B. Smith's *Spinoza's Book of Life: Freedom and Redemption in the Ethics* (New Haven: Yale University Press, 2003). This view is shared to a lesser extent by Jonathan I. Israel in his *Radical Enlightenment: Philosophy and the Making of Modernity 1650–1750* (Oxford: Oxford University Press, 2001).
41. Nancy K. Levene, *Spinoza's Revelation: Religion, Democracy, and Reason* (Cambridge: Cambridge University Press, 2004), pp. 10–12.
42. Benedict de Spinoza, "Ethics," in *A Spinoza Reader: The* Ethics *and Other Works*, ed. and trans. Edwin Curley (Princeton: Princeton University Press, 1994), IVPref.
43. Ibid., ID3.
44. Ibid., ID4.
45. Ibid., IIP13. It isn't entirely clear from the text why we can only have knowledge of these two and no more, though this doesn't seem really to be at issue for our understanding of the process of his thought.
46. Ibids., ID5.
47. This is the argument of Deleuze who places Spinoza among other authors steeped in Neoplatonism and developing various theories of expression and explication. He writes that the goal of these thinkers "was to thoroughly transform such Neoplatonism, to open it up to quite new lines of development, far removed from that of emanation, even where the two themes were both present" (Gilles Deleuze, *Expressionism in Philosophy: Spinoza*, trans. Martin Joughlin [New York: Zone Books, 1990], p. 19).
48. Levene, *Spinoza's Revelation*, p. 3.
49. Blayton Polka, *Between Philosophy and Religion: Spinoza, the Bible, and Modernity*, Vol. I: *Hermeneutics and Ontology* (Lexington: Lexington Books, 2007), p. 6.
50. Deleuze, *Expressionism in Philosophy*, p. 108.
51. Ibid., p. 109.
52. Ibid., p. 107.
53. Spinoza, *Ethics*, IIP7S.
54. Deleuze, *Expressionism in Philosophy*, p. 109.
55. Spinoza, *Ethics*, IA4.
56. Ibid., IP23.
57. See ibid., IP9.
58. Ibid., IIP7.

NOTES   255

59. Deleuze, *Expressionism in Philosophy*, p. 110.
60. Ibid.
61. Spinoza, *Theological-Political Treatise*, translated by Michael Silverthorne and Jonathan Israel (Cambridge: Cambridge University Press, 2007), p. 59.
62. Ibid.
63. IVP18S.
64. Conor Cunningham, *Genealogy of Nihilism: Philosophies of Nothing and the Difference of Theology* (London and New York: Routledge, 2002), pp. 60, 68.
65. See Jürgen Mettepennigen, *Nouvelle Théologie—New Theology: Inheritor of Modernism, Precursor of Vatican II* (London and New York: T&T Clark, 2010); and John Milbank, *The Suspended Middle: Henry de Lubac and the Debate Concerning the Supernatural* (London: SCM Press, 2005).
66. Henri de Lubac, *The Mystery of the Supernatural*, trans. Rosemary Sheed (New York: Herder and Herder, 1998) p. 63.
67. Spinoza, *Theological-Political Treatise*, p. 83.
68. Ibid., p. 81.
69. Deleuze, *Expressionism in Philosophy*, pp. 259–260.
70. Ibid., p. 259.
71. Spinoza, *Theological-Political Treatise*, p. 246.
72. Deleuze, *Expressionism in Philosophy*, p. 260.
73. Spinoza, *Theological-Political Treatise*, p. 197.
74. Deleuze, *Expressionism in Philosophy*, pp. 260–261.
75. Spinoza, *Theological-Political Treatise*, p. 241.
76. Ibid., p. 242.
77. Étienne Balibar, *Spinoza and Politics*, trans. Peter Snowdon (London and New York: Verso, 2008), pp. 7–8.
78. Spinoza, *Ethics*, VP25Dem.
79. See Alexandre Matheron, *Le Christ et le salut des ignorants chez Spinoza* (Paris: Aubier-Montaigne, 1971), for an extensive study of this aspect of Spinoza's thought.
80. François Laruelle, "The Decline of Materialism in the Name of Matter," trans. Ray Brassier, *Pli* 12 (2001): 37. This is a translation of a section from François Laruelle, *Le Principe de minoritié* (Paris: Aubier, 1981).
81. Cf. Laruelle, "The Decline of Materialism in the Name of Matter," p. 41.
82. Ya'qûb al-Sijistânî, *Le Dévoilement des choses cachées* (Kashf al-Mahjûb), trans. Henry Corbin (Lagrasse: Verdier, 1988), pp. 33–35.
83. Ibid., p. 81.
84. Christian Jambet, *La grande résurrection d'Alamût. Les formes de la liberté dans le shî'isme ismaélien* (Lagresse: Verdier, 1990), p. 210.
85. A summary of that formalism: There is the radical immanence of the One as One-in-One, which produces a philosophical resistance identified by Laruelle as non-thetic transcendence or non(-One) and a form of relative philosophical immanence as (non-)One. Both are effects of the One-in-One or Real.
86. See ibid., pp. 33–49, for a compressed history.

256   Notes

87. Christian Jambet, "The Paradoxical One," trans. Michael Stanish, *Umbr(a): A Journal of the Unconcious* (2009): 141; Jambet, *La Grande résurrection d'Alamût*, p. 142 [translation slightly modified].
88. Jambet, "The Paradoxical One," p. 142; *La Grande résurrection*, p. 143.
89. al-Sijistânî, *Le Dévoilement des choses caches*, p. 77.
90. Christian Jambet and Guy Lardreau, *L'Ange. Pour une cynégétique du semblant, Ontologie de la révolution 1* (Paris: Grasset, 1976), p. 36.
91. Cf. ibid., pp. 213–224.
92. Cf. ibid., p. 92.
93. Ibid.
94. Ibid., p. 84.
95. Ibid., p. 87.
96. Ibid., p. 100.
97. Ibid., p. 109.
98. Ibid., p. 20.
99. Of course, it may also give birth to the barbaric Angel. See Christian Jambet and Guy Lardreau, *Le Monde. Reponse à la question: Qu'est-ce que les droits de l'homme?* (Paris: Grasset, 1978), p. 187. Lardreau is more direct about this in his own work of negative philosophy entitled *La véracité*, where he argues for a Kantian sublime within politics defined as "a politics that makes a finality sensible to us that is completely independent from nature" (Guy Lardreau, *La véractié. Essai d'une philosophie négative* [Lagrasse: Verdier, 1993], p. 237). Lardreau again invokes the Angel in his development of the concept of the political sublime, this time as the "political name for the desire for death" (p. 241). Within a negative philosophy this desire for death is limited, it is a desire for the self-referential play of the correlative images of the self and the other. In the terms laid out in my "The Judgement of God and the Immeasurable" it is the desire for the death of the play between friend and enemy. For Lardreau, within a negative philosophy, this desire is checked by way of a negative presentation of the Real (ibid). Death is always a form of transcendence as limit for philosophers and Lardreau is no different (Lardreau, *La véractié*, p. 243. Cf. Philip Goodchild, *Capitalism and Religion: The Price of Piety* (London and New York: Routledge, 2002), pp. 148–155). The barbarous Angel, for Lardreau, comes when there is a positive presentation of the Real, a presentation that threatens to topple the sublime over (Lardreau, *La véractié*, p. 241). See Anthony Paul Smith, "Nature Deserves to Be Side by Side with the Angels: Nature and Messianism by way of Non-Islam," in *Angelaki* (Forthcoming, 2014) for a longer discussion of this angelology.
100. Jambet, *Le Grande resurrection*, p. 362.
101. Francois Laruelle, *Future Christ: A Lesson in Heresy*, trans. Anthony Paul Smith (London and New York: Continuum, 2010), p. 17; *Le Christ futur. Une leçon d'hérésie* (Paris: Exils Éditeur, 2002), p. 33.
102. Laurelle, *Future Christ*, pp. 39, 40; *Le Christ futur*, pp. 58, 59.
103. Laruelle, *Future Christ*, p. 6; *Le Christ futur*, p. 20.
104. Laruelle, *Future Christ*, p. 117; *Le Christ futur*, p. 145.

NOTES   257

105. François Laruelle, "A Science in (*en*) Christ," trans. Aaron Riches, in *The Grandeur of Reason: Religion, Tradition and Universalism*, eds. Peter M. Candler, Jr. and Conor Cunningham (London: SCM Press, 2010), p. 318.

106. Gilles Grelet, *Déclarer la gnose. D'une guerre qui revient à la culture* (Paris: L'Harmattan, 2002), pp. 119f50, 119f49.

107. Interestingly al-Sijistânî develops an Islamic Christology by way of a kind of fourfold, since he uses the image of the cross as a hermeneutic for understanding Christ. See Abu Ya'qûb al-Sijistânî, "The Book of Wellsprings," in *The Wellsprings of Wisdom*, ed. and trans. Paul E. Walker (Salt Lake City: University of Utah Press, 1994), pp. 37–111. In the editors' introduction to *After the Postsecular and the Postmodern*, Daniel Whistler and I differentiated between a postsecular *event* and the appropriation of that event in the name of a theologization of philosophy or what we called "imperial secularism." The event marked a break with Western imperialism, which used Christian forms of thought to develop a post-Christian secularism in an attempt to separate the oppressed colonial subjects internally—a separation of the political and their religious identity, whereas the appropriation of the event is often an attempt to reinstate (at best) a war at the ideational level and (at worst) a new form of imperial war in the name of the clash of traditions. The postsecular event, we claim there, was located largely with Islamic countries throughout the Middle East and North Africa and though the response to the postsecular event is not an Islamic turn in Continental philosophy, there should be more engagement with forms of thought outside of the Christian tradition. There are two clear non-theological reasons for this: (1) if the generic is to be located in a way that avoids the shortcomings of Hegelian philosophy and its continuing influence on the practice of philosophy of religion, where European Christianity comes to be the name for universalism as the only consummating historical religion, then it must take the infinite task of working with any material whatsoever in order to locate the power of the generic that lies there; (2) non-theology always begins from the perspective of the murdered, and thus from the perspective of heretical material, which is to say that there is within non-theology a principle of minority or preferential option for the poor as immanent to generic humanity. With regard to the second, in a very real sense, a very bodily sense, a certain appearance of the power of poverty, what Negri calls the "force of the slave" in regard to Job, has coalesced around the name "Muslim" (though, of course, not *just* the name "Muslim"). In Europe the Muslim has become the exception that grounds the law, both political law and the economic law of class difference. As this structural aspect in-person the Muslim is, as Mehdi Belhaj Kacem has argued, the contemporary form of pariah: "The pariah is at once captured and delivered, locked within its exclusion and banished by inclusion" (Mehdi Belhaj Kacem, *La psychose française. Les banlieues: le ban de la République* [Paris: Gallimard, 2006], p. 18). The reality of the pariah is manifested clearly in the collusion of the institutional Left with the establishment Right of Europe regarding these "places of the ban" (Belhaj Kacem makes a clever play on the name of the suburban ghettoes of France,

258 NOTES

les banlieus, as *les ban-lieux*) as *problems* to be neutralized (and both speak in this language if with differing degrees of violence) while also referring to them as what negatively grounds their existence as government.

Islam as pariah is repeated within thought in the same way as it is found in the political field. Within a teleological-oriented philosophy of religion, which always carries with it a certain amount of servitude to the European project of expansion, Islam poses a problem because it arises historically after the Christianity, which was to be the consummation of Spirit in the religious realm. See Alberto Toscano, *Fanaticism: On the Uses of an Idea* (London: Verso, 2010), p. 164, and the entirety of the chapter "The Revolution of the East," which is an excellent examination and exposition of the weaknesses of the engagement of Hegel and Žižek with Islam, specifically with regard to the One of Islam. Islam is the religion that proves Christianity's true universalism, as an idea made concrete, while (as Toscano says) Islam "takes universalism too far" in its abstract passion for the real (p. 153). As such, it is both brought within the scope of the historical development of European spirit and excluded from it as a form of fanaticism that threatens the rational integration of people within the State. There is a ban on Islam in Continental philosophy, known for its engagement with Jewish and Christian religious thought, that grounds its practice within post-Christian secularism insofar as Islam is so dangerous as to be outside this form of the secular. This is why Islam (though, of course, not just Islam; we're not arguing here for a simple "Islamic turn") must be material for the construction of a true non-theology, not in the name of an Islamic theologization of philosophy, but because the event taking place among Muslims, both as postsecular event and as name of the pariah, have consequences for philosophy and may help free philosophy, practiced here as non-theology, from its capture as imperialist weapon.

CONCLUSION: THEORY OF NATURE

1. St. Thomas Aquinas, *Summa Contra Gentiles*, trans. A. C. Pegis, James F. Anderson, Vernon J. Bourke, and Charles J. O'Neil, 4 vols. (Notre Dame, IN: University of Notre Dame Press, 1975), II/2.2; hereafter SCG.
2. SCG II/2.3.
3. SCG II/2.4.
4. SCG II/2.5.
5. Eugene F. Rogers, *Thomas Aquinas and Karl Barth: Sacred Doctrine and the Natural Knowledge of God* (Notre Dame and London: University of Notre Dame Press, 1995), p. 5.
6. For a helpful discussion of this separation from what could be termed a non-theological perspective, see Eugene Thacker, *After Life* (Chicago: University of Chicago Press, 2010), pp. 104–107.

NOTES   259

7. See Roger D. Sorrel, *St. Francis of Assisi and Nature: Tradition and Innovation in Western Christian Attitudes towards the Environment* (Oxford: Oxford University Press, 1988), pp. 55–97.
8. Brethren of Purity, *Epistles of the Brethren of Purity: The Case of the Animals versus Man before the King of the Jinn*, eds. and trans. Lenn E. Goodman and Richard McGregor (Oxford: Oxford University Press, 2009), p. 102.
9. IVP39S and see François Zourabichvili, *Le conservatisme paradoxal de Spinoza. Enfance et royauté* (Paris: PUF, 2002), pp. 95–177.
10. François Zourabichvili, *Spinoza. Une physique de la pensée* (Paris: PUF, 2002), p. 220; my emphasis.
11. François Laruelle, *Philosophies of Difference: A Critical Introduction to Non-Philosophy*, trans. Rocco Gangle (London and New York: Continuum, 2010), p. 202; *Les philosophies de la différence* (Paris: PUF, 1986), p. 219.
12. Friedrich Nietzsche, *Thus Spoke Zarathustra*, trans. Walter Kaufmann (New York: Modern Library, 1995), p. 13.
13. Christian Jambet, "A Philosophical Commentary," trans. Hafiz Karmali, in *Paradise of Submission: A Medieval Treatise on Ismaili Thought*, ed. and trans. S. J. Badakhchani (London and New York: I.B Tauris, 2005), p. 181.
14. Naṣīr al-Dīn Ṭūsī, *Paradise of Submission: A Medieval Treatise on Ismaili Thought*, ed. and trans. S. J. Badakhchani (London and New York: I.B Tauris, 2005), p. 16.
15. Jambet, "A Philosophical Commentary," 181.
16. Cf. ibid., p. 178.

Bibliography

Agamben, Giorgio. *The Open: Man and Animal.* Translated by Kevin Attell. Stanford: Stanford University Press, 2004.

al-Dīn Ṭūsī, Naṣīr. *Paradise of Submission: A Medieval Treatise on Ismaili Thought.* Edited and translated by S. J. Badakhchani. London and New York: I.B Tauris, 2005.

al-Sijistânî, Ya'qûb. "The Book of Wellsprings," in *The Wellsprings of Wisdom.* Edited and translated by Paul E. Walker. Salt Lake City: University of Utah Press, 1994.

———. *Le Dévoilement des choses cachées* (Kashf al-Mahjûb). Translated by Henry Corbin. Lagrasse: Verdier, 1988.

Althusser, Louis. *For Marx.* Translated by Ben Brewster. London: Verso, 2005.

Aquinas, St. Thomas. "Commentary on Boethius' *De Trinitate* [Selection]," in *Selected Philosophical Writings.* Edited and translated by Timothy McDermott. Oxford: Oxford University Press, 1993.

———. "On Kingship or the Governance of Rulers (De Regimine Principum, 1265–1267)," in *St. Thomas Aquinas on Politics and Ethics.* Edited and translated by Paul E. Sigmund. London and New York: W. W. Norton & Company, 1988.

———. "On the Principles of Nature," in *Selected Philosophical Writings.* Edited and translated by Timothy McDermott. Oxford: Oxford University Press, 1993.

———. *Summa Contra Gentiles.* Translated by A. C. Pegis, James F. Anderson, Vernon J. Bourke, and Charles J. O'Neil. 4 Volumes. Notre Dame, IN: University of Notre Dame Press, 1975.

———. *Summa Theologica (literal English translation).* Translated by the Fathers of the English Dominican Province. Westminster, MD: Christian Classics, 1981.

Asad, Talal. *Formations of the Secular: Christianity, Islam, Modernity.* Stanford: Stanford University Press, 2003.

Badiou, Alain. *Being and Event.* Translated by Oliver Feltham. New York and London: Continuum, 2005.

———. *The Century.* Translated by Alberto Toscano. Cambridge: Polity, 2007.

———. Interview with Ben Woodward. *The Speculative Turn: Continental Materialism and Realism.* Edited by Levi Bryant, Nick Srnicek, and Graham Harman. Melbourne: re.press, 2011.

———. *Logics of Worlds: Being and Event II.* Translated by Alberto Toscano. New York and London: Continuum, 2009.

———. *Manifeste pour la philosophie.* Paris: Éditions du Seuil, 1980.

262 BIBLIOGRAPHY

Badiou, Alain. "Mathematics and Philosophy: The Grand Style and the Little Style," in *Theoretical Writings*. Edited and translated by Ray Brassier and Alberto Toscano. London and New York: Continuum, 2004.

Baerlocher, Felix. "The Gaia Hypothesis—A Fruitful Fallacy." *Experientia* 46.3 (1990): 232–238.

Balibar, Étienne. *Spinoza and Politics*. Translated by Peter Snowdon. London and New York: Verso, 2008.

Barber, Daniel Colucciello. *Deleuze and the Naming of God: Post-Secularism and the Future of Immanence*. Edinburgh: Edinburgh University Press, forthcoming.

Barth, Karl. *Church Dogmatics* III.4. Translated by A. T. Mackay, T. H. L. Parker, H. Knight, H. A. Kennedy, and J. Marks. Edited by G. W. Bromiley and T. F. Torrance. London and New York: T&T Clark, 2004.

Belhaj Kacem, Mehdi. *La psychose françiase. Les banlieues: le ban de la République*. Paris: Gallimard, 2006.

Bell, Jr., Daniel M. "Only Jesus Saves: Toward a Theo-political Ontology of Judgment," in *Theology and the Political: The New Debate*. Edited by Creston Davis, John Milbank, and Slavoj Žižek. Durham: Duke University Press, 2005.

Benedict XVI, Pope. *Encyclical Letter, Caritas in veritate: Charity in Love*. Rome: St. Peter's, 2009. Available online: http://www.vatican.va/holy_father/benedict_xvi /encyclicals/documents/hf_ben-xvI_enc_20090629_caritas-in-veritate_En.html (accessed June 13, 2010).

Boff, Leonardo. *Cry of the Earth, Cry of the Poor*. Translated by Philip Berryman. Maryknoll, NY: Orbis Books, 1997.

———. *Ecology and Liberation: A New Paradigm*. Translated by John Cumming. Maryknoll, NY: Orbis Books, 1995.

Botkin, Daniel B. *Discordant Harmonies: A New Ecology for the Twenty-First Century*. Oxford: Oxford University Press, 1990.

Brassier, Ray. *Nihil Unbound: Enlightenment and Extinction*. Basingstoke: Palgrave Macmillan, 2007.

Brassier, Ray, Iain Hamilton Grant, Graham Harman, and Quentin Meillassoux. "Speculative Realism," in *Collapse III*. Falmouth: Urbanomic, 2007: 307–449.

Brethren of Purity. *Epistles of the Brethren of Purity: The Case of the Animals versus Man Before the King of the Jinn*. Edited and translated by Lenn E. Goodman and Richard McGregor. Oxford: Oxford University Press, 2009.

Brown, Charles S., and Ted Toadvine, editors. *Eco-Phenomenology: Back to the Earth Itself*. Albany: State University of New York Press, 2003.

———. *Nature's Edge: Boundary Explorations in Ecological Theory and Practice*. Albany: State University of New York Press, 2007.

Buchanan, Brett. *Onto-Ethologies: The Animal Environments of Uexküll, Heidegger, Merleau-Ponty, and Deleuze*. Albany: SUNY Press, 2008.

Bufalo, Erik del. *Deleuze et Laruelle. De la schizo-analyse à la non-philosophie*. Paris: Kimé, 2003.

Chapin III, F. Stuart, Pamela A. Matson, and Harold A. Mooney. *Principles of Terrestrial Ecosystem Ecology*. London: Springer, 2002.

Choplin, Hugues. *De la phénoménologie à la non-philosophie. Lévinas et Laruelle*. Paris: Kimé, 1997.

BIBLIOGRAPHY   263

——. *L'espace de la pensée française contemporaine.* Paris: L'Harmattan, 2007.
Coates, Peter. *Nature: Western Attitudes since Ancient Times.* Cambridge: Polity
    Press, 1998.
Corbin, Henry. *Le Paradoxe du Monothéisme.* Paris: L'Herne, 2003.
Crosby, Alfred W. *Ecological Imperialism: The Biological Expansion of Europe, 900–*
    1900. Cambridge: Cambridge University Press, 1986.
Cunningham, Conor. *Genealogy of Nihilism: Philosophies of Nothing and the*
    Difference of Theology. London and New York: Routledge, 2002.
Deleuze, Gilles. *Bergsonism.* Translated by Hugh Tomlinson and Barbara
    Habberjam. New York: Zone Books, 1991.
——. *Difference and Repetition.* Translated by Paul Patton. New York: Columbia
    University Press, 1995.
——. *Expressionism in Philosophy: Spinoza.* Translated by Martin Joughlin. New
    York: Zone Books, 1990.
Deleuze, Gilles, and Félix Guattari. *What is Philosophy?* Translated by Hugh
    Tomlinson and Graham Burchell. New York: Columbia University Press,
    1994.
Diamond, Jared. *Collapse: How Societies Choose to Succeed or Fail.* London and
    New York: Penguin, 2006.
——. *Guns, Germs and Steel: A Short History for the Last 13,000 Years.* London:
    Vintage, 2005.
Fink, Eugen. *Sixth Cartesian Mediation: The Idea of a Transcendental Theory of*
    Method. Translated by Ronald Bruzina. Bloomington: Indiana University
    Press, 1995.
Foltz, Bruce V. *Inhabiting the Earth: Heidegger, Environmental Ethics, and the*
    Metaphysics of Nature. New Jersey: Humanities Books, 1996.
Fontaine, Patrick. *Platon autrement dit.* Paris: L'Harmattan, 2007.
Fraser, Zachery Luke. Draft of "Entry for 'Generic,'" in *The Badiou Dictionary.*
    Edinburgh: Edinburgh University Press, forthcoming. Available online:
    http://formandformalism.blogspot.com/2011/03/generic-entry.html (accessed
    March 25, 2011).
Frodemen, Robert. *Geo-Logic: Breaking Ground Between Philosophy and the Earth*
    Sciences. Albany: State University of New York Press, 2003.
Gangle, Rocco. "Theology of the Chimera: Spinoza, Immanence, Practice,"
    in *After the Postsecular and the Postmodern: New Essays in Continental*
    Philosophy of Religion. Edited by Anthony Paul Smith and Daniel Whistler.
    Newcastle-upon-Tyne: Cambridge Scholars' Press, 2010.
Giller, Paul S. *Community Structure and the Niche.* London and New York:
    Chapman and Hall, 1984.
Golley, Frank B. *A History of the Ecosystem Concept in Ecology: More Than the Sum*
    of the Parts. New Haven and London: Yale University Press, 1993.
Goodchild, Philip. *Theology of Money.* Durham & London: Duke University
    Press, 2009.
——. *Capitalism and Religion: The Price of Piety.* London: Routledge, 2002.
——. "Oil and Debt—the Collision between Ecology and Economy." *Situation*
    Analysis 2 (Spring 2003): 5–18.

Grant, Iain Hamilton. "The 'Eternal and Necessary Bond between Philosophy and Physics': A Repetition of the Difference between the Fichtean and Schellingian Systems of Philosophy." *Angelaki: Journal of the Theoretical Humanities* 10:1 (April 2005): 43–59.
———. "The Insufficiency of Ground: On Žižek's Schellingianism," in *The Truth of Žižek*. Edited by Paul Bowman and Richard Stamp. London and New York: Continuum, 2007.
———. *Philosophies of Nature after Schelling*. London and New York: Continuum, 2006.
Grelet, Gilles. *Déclarer la gnose. D'une guerre qui revient à la culture*. Paris: L'Harmattan, 2002.
Hadot, Pierre. *The Veil of Isis: An Essay on the History of the Idea of Nature*. Translated by Michael Chase. London and Cambridge, MA: The Belknap Press of Harvard University Press, 2006.
Harlingue, Olivier. *Sans condition. Blanchot, la littérature, la philosophie*. Paris: L'Harmattan, 2009.
Harman, Graham. *Prince of Networks: Bruno Latour and Metaphysics*. Melbourne: re:press, 2009.
———. *The Quadruple Object*. Winchester, UK: Zero Books, 2011.
Heidegger, Martin. *Being and Time*. Translated by John Macquarrie and Edward Robinson. New York: HarperSanFrancisco, 1962.
———. *The Fundamental Concepts of Metaphysics: World, Finitude, Solitude*. Translated by William McNeill and Nicholas Walker. Bloomington: Indiana University Press, 1995.
———. "Modern Science, Metaphysics, and Mathematics," in *Basic Writings*. Edited and translated by David Farrell Krell. New York: HarperSanFrancisco, 1993.
———. "…Poetically Man Dwells…," in *Poetry, Language, Thought*. Edited and translated by Albert Hofstadter. New York: Harper Perennial Modern Classics, 1971.
———. "The Question Concerning Technology," in *Basic Writings*. Edited and translated by David Farrell Krell. San Francisco: HarperSanFrancisco, 1993.
———. "The Thing," in *Poetry, Language, Thought*. Translated and edited by Albert Hofstadter. New York: Harper Perennial Modern Classics, 1971.
Henry, Michel. *I Am the Truth: Toward a Philosophy of Christianity*. Translated by Susan Emmanuel. Stanford: Stanford University Press, 2003.
Husserl, Edmund. *The Crisis of European Sciences and Transcendental Phenomenology: An Introduction to Phenomenological Philosophy*. Translated by David Carr. Evanston: Northwestern University Press, 1970.
Israel, Jonathan I. *Radical Enlightenment: Philosophy and the Making of Modernity 1650–1750*. Oxford: Oxford University Press, 2001.
Jambet, Christian. *La grande résurrection d'Alamût. Les formes de la liberté dans le shî'isme ismaélien*. Lagresse: Verdier, 1990.
———. "The Paradoxical One." Translated by Michael Stanish. *Umbr(a): A Journal of the Unconcious* (2009): 139–163.

BIBLIOGRAPHY 265

———. "A Philosophical Commentary." Translated by Hafiz Karmali in *Paradise of Submission: A Medieval Treatise on Ismaili Thought*. Edited by S. J. Badakhchani. London and New York: I.B Tauris, 2005.

Jambet, Christian, and Guy Lardreau. *L'Ange. Pour une cynégétique du semblant, Ontologie de la révolution 1*. Paris: Grasset, 1976.

———. *Le Monde. Reponse à la question: Qu'est-ce que les droits de l'homme?* Paris: Grasset, 1978.

Janicaud, Dominique. *Phenomenology "Wide Open": After the French Debate.* Translated by Charles N. Cabral. New York: Fordham University Press, 2005.

Jenkins, Willis. *Ecologies of Grace: Environmental Ethics and Christian Theology.* Oxford: Oxford University Press, 2008.

Jordan, Mark D. *Rewritten Theology: Aquinas after His Readers.* Oxford: Blackwell, 2006.

Jürgen Mettepennigen, *Nouvelle Théologie—New Theology: Inheritor of Modernism, Precursor of Vatican II.* London and New York: T&T Clark, 2010.

Katz, Eric. "The Big Lie: Human Restoration of Nature." *Research in Philosophy and Technology* 12 (1992): 231–241.

Kenny, Anthony. *Aquinas on Being.* Oxford: Oxford University Press, 2002.

Khaldûn, Ibn. *The Muqaddimah: An Introduction to History.* Edited and translated by Franz Rosenthal. Princeton: Princeton University Press, 1969.

Kieffer, Gilbert. *Esthétiques non-philosophiques.* Paris: Kimé, 1996.

———. *Que peut la peinture pour l'esthétique?* Paris: Éditions Petra, 2003.

Lardreau, Guy. *La véractié. Essai d'une philosophie negative.* Lagrasse: Verdier, 1993.

Laruelle, François. *Anti-Badiou: On the Introduction of Maoism into Philosophy.* Translated by Robin Mackay. London and New York: Bloomsbury, 2013.

———. "The Decline of Materialism in the Name of Matter." Translated by Ray Brassier. *Pli* 12 (2001): 33–40.

———. "The Degrowth of Philosophy: Toward a Generic Ecology," in *From Decision to Heresy: Experiments in Non-Standard Thought*. Edited and translated by Robin Mackay. Falmouth: Urbanomic, 2013: 327–349.

———. *En tant qu'Un: La « non-philosophie » expliquée aux philosophes.* Paris: Aubier, 1991.

———. "From the First to the Second Non-Philosophy," in *From Decision to Heresy: Experiments in Non-Standard Thought*. Edited by Robin Mackay. Translated by Anthony Paul Smith and Nicola Rubczak. Falmouth: Urbanomic, 2013: 305–325.

———. *Introduction au non-marxisme.* Paris: PUF, 2000.

———. *Introduction aux sciences generiques.* Paris: Pétra, 2008.

———. *La Lutte et l'Utopie à la fin des temps philosophiques.* Paris: Kimé, 2004. [*Struggle and Utopia at the End Times of Philosophy*. Translated by Drew S. Burk and Anthony Paul Smith. Minneapolis: Univocal Publishing, 2012.]

———. *Le Christ futur. Une leçon d'hérèsie.* Paris: Exils Éditeur, 2002. [*Future Christ: A Lesson in Heresy*. Translated by Anthony Paul Smith. London and New York: Continuum, 2010.]

———. *Le Principe de minorité.* Paris: Aubier, 1981.

266 BIBLIOGRAPHY

Laruelle, François. *Les philosophies de la différence. Introduction critique.* Paris: PUF, 1986. [*Philosophies of Difference: A Critical Introduction to Non-Philosophy.* Translated by Rocco Gangle. London and New York: Continuum, 2010.]

———. "L'Impossible foundation d'une écologie de l'océan." Available online: http://www.onphi.net/lettre-laruelle-l-impossible-fondation-d-une-ecologie -de-l-ocean-27.html (accessed March 22, 2011).

———. *Mystique non-philosophique à l'usage des contemporains.* Paris: L'Harmattan, 2007.

———. "A New Presentation of Non-Philosophy." Working paper for L'Organisation Non-Philosophique Internationale. February 11, 2004. Available online: http://www.onphi.net/texte-a-new-presentation-of-non-philosophy-32.html (accessed February 1, 2010).

———. *Philosophie et non-philosophie.* Mardaga: Liege-Bruxelles, 1989.

———. *Philosophie non-standard. Générique, quantique, philo-fiction.* Paris: Kimé, 2010.

———. *Principes de la non-philosophie.* Paris: PUF, 1995. [*Principles of Non-Philosophy.* Translated by Nicola Rubczak and Anthony Paul Smith. London and New York: Bloomsbury, 2013.]

———. "A Science in [*en*] Christ," in *The Grandeur of Reason: Religion, Tradition and Universalism.* Edited by Peter M. Candler, Jr. and Conor Cunningham. Translated by Aaron Riches. London: SCM Press, 2010: 316–331.

———. *Théorie des identités. Fractalité généralisée et philosophie artificielle.* Paris: PUF, 1992.

———. *Une Biographie de l'homme ordinaire. Des Autorités et des Minorités.* Paris: Aubier, 1985.

Laruelle, François, et al. *Dictionnaire de la non-philosophie.* Paris: Kimé, 1998.

Latour, Bruno. *Politics of Nature: How to Bring the Sciences into Democracy.* Translated by Catherine Porter. Cambridge, MA, and London: Harvard University Press, 2004.

Levene, Nancy K. *Spinoza's Revelation: Religion, Democracy, and Reason.* Cambridge: Cambridge University Press, 2004.

Leopold, Aldo. *A Sand County Almanac.* Oxford: Oxford University Press, 1968.

Lévêque, Christian. *Ecology: From Ecosystem to Biosphere.* Plymouth, UK: Science Publisher, Inc., 2003.

Lewis, Michael. *Heidegger beyond Deconstruction: On Nature.* London and New York: Continuum, 2007.

Llewelyn, John. *The Middle Voice of Ecological Conscience.* Basingstoke: Macmillan, 1991.

———. *Seeing through God: A Geophenomenology.* Bloomington and Indianapolis: Indiana University Press, 2004.

Lubac, Henri de. *The Mystery of the Supernatural.* Translated by Rosemary Sheed. New York: Herder and Herder, 1998.

Marx, Karl. "Theses on Feuerbach," in *The Marx and Engels Reader.* Edited and translated by Robert C. Tucker. New York and London: W. W Norton & Company, 1978.

Matheron, Alexandre. *Le Christ et le salut des ignorants chez Spinoza.* Paris: Aubier-Montaigne, 1971.

McFague, Sallie. *The Body of God: An Ecological Theology.* London: SCM Press, 1993.

Meillassoux, Quentin. *After Finitude: An Essay on the Necessity of Contingency.* Translated by Ray Brassier. London and New York: Continuum, 2008.

Milbank, John. "Knowledge: The Theological Critique of Philosophy in Hamann and Jacobi," in *Radical Orthodoxy: A New Theology.* Edited by John Milbank, Catherine Pickstock, and Graham Ward. London: Routledge, 1999.

———. *The Suspended Middle: Henry de Lubac and the Debate Concerning the Supernatural.* London: SCM Press, 2005.

Milbank, John, and Catherine Pickstock. "Truth and Correspondence," in *Truth in Aquinas.* London: Routledge, 2001.

Mollet, Eric. *Bourdieu et Laruelle. Sociologie réflexive et non-philosophie.* Paris: Éditions Petra, 2003.

Montagnes, Bernard. *The Doctrine of the Analogy of Being According to Thomas Aquinas.* Translated by E. M. Macierowski. Milwaukee, WI: Marquette University Press, 2004.

Morton, Timothy. *The Ecological Thought.* London and Cambridge, MA: Harvard University Press, 2010.

———. *Ecology without Nature: Rethinking Environmental Aesthetics.* Cambridge, MA, and London: Harvard University Press, 2007.

Moulinier, Didier. *De la psychanalyse à la non-philosophie. Lacan et Laruelle.* Paris: Kimé, 1998.

Mullarkey, John. "A Bellicose Democracy: Bergson on the Open Soul (or Unthinking the Thought of Equality)," in *After the Postsecular and the Postmodern: New Essays in Continental Philosophy of Religion.* Edited by Anthony Paul Smith and Daniel Whistler. Newcastle-upon-Tyne: Cambridge Scholars Publishing, 2010.

———. *Refractions of Reality: Philosophy and the Moving Image.* Basingstoke: Palgrave, 2008.

———. *Post-Continental Philosophy: An Outline.* London and New York: Routledge, 2006.

Nadler, Steven. *Spinoza: A Life.* Cambridge: Cambridge University Press, 1999.

Naess, Arne. *Ecology, Community and Lifestyle: Outline of an Ecosophy.* Translated by David Rothenberg. Cambridge: Cambridge University Press, 1989.

———. "Spinoza and Ecology." *Philosophia* 7:1 (March 1977): 45–54.

Negri, Antonio. *The Labor of Job: The Biblical Text as a Parable for Human Labor.* Translated by Matteo Mandarini. Durham and London: Duke University Press, 2009.

Nietzsche, Friedrich. *Thus Spoke Zarathustra.* Translated by Walter Kaufmann. New York: Modern Library, 1995.

Northcott, Michael S. *The Environment and Christian Ethics.* Cambridge: Cambridge University Press, 1996.

Polka, Blayton. *Between Philosophy and Religion: Spinoza, the Bible, and Modernity.* Vol. I: *Hermeneutics and Ontology.* Lexington: Lexington Books, 2007.

Primavesi, Anne. *Sacred Gaia: Holistic Theology and Earth System Science.* London and New York: Routledge, 2000.

Radkau, Joachim. *Nature and Power: A Global History of the Environment.* Translated by Thomas Dunlap. Cambridge: Cambridge University Press, 2008.

Rannou, Jean-Luc. *La non-philosophie, simplement. Une introduction synthétique.* Paris: L'Harmattan, 2005.

Rogers, Eugene F. *Thomas Aquinas and Karl Barth: Sacred Doctrine and the Natural Knowledge of God.* Notre Dame and London: University of Notre Dame Press, 1995.

Ruether, Rosemary Radford. *Gaia & God: An Ecofeminist Theology of Earth Healing.* London: SCM Press, 1992.

Schmid, Anne-Françoise. "Le problème de Russell," in *La Non-philosophe des contemporains.* Paris: Kimé, 1995.

Smith, Anthony Paul. "Believing in this World for the Making of Gods: On the Ecology of the Virtual and the Actual." *SubStance* 38.3 (April 2010): 101–112.

———. "Philosophy and Ecosystem: Towards a Transcendental Ecology." *Polygraph* 22 (2010): 65–82.

———. "What Can Be Done with Religion?: Non-Philosophy and the Future of Philosophy of Religion," in *After the Postsecular and the Postmodern: New Essays in Continental Philosophy of Religion.* Edited by Anthony Paul Smith and Daniel Whistler. Newcastle-upon-Tyne: Cambridge Scholars Publishing, 2010.

Smith, Stephen B. *Spinoza's Book of Life: Freedom and Redemption in the Ethics.* New Haven: Yale University Press, 2003.

Sorrel, Roger D. *St. Francis of Assisi and Nature: Tradition and Innovation in Western Christian Attitudes towards the Environment.* Oxford: Oxford University Press, 1988.

Spinoza, Benedict de. "Ethics," in *A Spinoza Reader: The Ethics and Other Works.* Edited and translated by Edwin Curley. Princeton: Princeton University Press, 1994.

———. *Theological-Political Treatise.* Translated by Michael Silverthorne and Jonathan Israel. Cambridge: Cambridge University Press, 2007.

Tansley, A. G. "The Use and Abuse of Vegetational Concepts and Terms." *Ecology* 16.3 (1935): 299–303.

Taylor, Paul W. *Respect for Nature: A Theory of Environmental Ethics.* Princeton, NJ: Princeton University Press, 1986.

Thacker, Eugene. *After Life.* Chicago, University of Chicago Press, 2010.

Toadvine, Ted. *Merleau-Ponty's Philosophy of Nature.* Evanston: Northwestern University Press, 2009.

Torrell O. P., Jean-Pierre. *Saint Thomas Aquinas.* Translated by Robert Royal. Vol. I: *The Person and His Work.* Washington, D.C.: Catholic University of America Press, 1996.

———. *Saint Thomas Aquinas.* Translated by Robert Royal. Vol. II: *Spiritual Master.* Washington D.C.: Catholic University of America Press, 2003.

Toscano, Alberto. *Fanaticism: On the Uses of an Idea.* London: Verso, 2010.

BIBLIOGRAPHY   269

Velde, Rudi te. *Aquinas on God: The "Divine Science" of the Summa theologiae.*
  Aldershot: Ashgate, 2006.
Walker, Brian, and David Salt. *Resilience Thinking: Sustaining Ecosystems and
  People in a Changing World.* Washington and London: Island Press, 2006.
Whistler, Daniel. "Language after Philosophy of Nature: Schelling's Geology
  of Divine Names," in *After the Postsecular and the Postmodern: New Essays in
  Continental Philosophy of Religion.* Edited by Anthony Paul Smith and Daniel
  Whistler. Newcastle-upon-Tyne: Cambridge Scholars Publishing, 2010.
Wilson, Edward O. *The Creation: An Appeal to Save Life on Earth.* New York:
  W. W. Norton & Company, 2006.
————. *The Diversity of Life.* London: Penguin Books, 2001.
Worster, Donald. *Nature's Economy: A History of Ecological Ideas.* 2nd Edition.
  Cambridge: Cambridge University Press, 1994.
Young, Julian. "The Fourfold" in *The Cambridge Companion to Heidegger.* Edited
  by Charles B Guignon. 2nd Edition. Cambridge: Cambridge University Press,
  2006.
Žižek, Slavoj. "Censorship Today: Violence, or Ecology as a New Opium for the
  Masses." Available online: http://www.lacan.com/zizecology1.htm (accessed
  September 10, 2010).
————. *In Defense of Lost Causes.* London and New York: Verso, 2008.
————. *The Indivisible Remainder: On Schelling and Related Matters.* London and
  New York: Verso, 2007.
Zourabichvili, François. *Le conservatisme paradoxal de Spinoza. Enfance et royauté.*
  Paris: PUF, 2002.
————. *Spinoza. Une physique de la pensée.* Paris: PUF, 2002.

Index

actuality 21, 35, 43, 77, 106, 126, 133, 136, 146, 178, 186, 196, 202, 206–8
aesthetics, environmental 27
alterity 2, 5–6, 10, 72, 76–8, 99, 113, 133, 164, 168, 170, 184–5, 211, 216, 219, 222
Althusser, Louis 88
analogia entis 171–3 *see also* analogy of being
analogy 153, 168, 191–8, 201, 219–20
analogy of being 171–2, 203
angels 210, 213, 215–16
antiphilosophy 66, 96
apophaticism 190, 192–3, 197, 219
significant 193, 197
Aquinas, St. Thomas 15–16, 47, 49, 113, 115, 171–5, 190–8, 202–4, 208–9, 218–19
attributes 200–2, 207, 212
authority 16, 48, 75, 79, 87, 97–8, 102–7, 110, 160–1, 171, 174, 177, 189, 205–6, 215
autonomy, radical 70, 79, 118, 147–8, 211
axioms 5–6, 14, 30, 43–4, 61, 69, 71, 75–6, 80–1, 85, 87, 91, 97, 99–100, 102–3, 109, 174, 182

Badiou, Alain 4, 25, 62–3, 66, 82–4, 91, 96–7, 109, 118, 168, 177, 179–83, 210
Barber, Daniel Colucciello 242f1
Barth, Karl 6, 47, 105

being
-generic 92
-in-the-world 120, 178
Benedict XVI, Pope 6, 8, 49, 53, 121, 174, 187
biodiversity 31, 42–3, 60, 113, 116, 131–2, 134–9, 145, 150, 219
biological diversity 131, 134
biology 26, 29, 36–7, 82, 125–6, 131
biosphere 1, 10, 20, 24–5, 32, 38, 48–9, 52, 54, 104, 115, 128, 134–7, 139, 143, 146–7, 151–2, 175, 185–7, 219–20
bodies 5, 51, 54, 92, 108, 143, 145, 149, 182–3, 201, 205–6, 214, 221, 223
Boff, Lenoardo 53–4, 56
Botkin, Daniel B. 20–5, 138
Brassier, Ray 33, 36, 39, 42, 63–5, 67, 71, 77, 148
Brethren of Purity 220

capitalism 3, 50, 79, 160, 213
carbon 146–7
causality 79–81, 89, 113, 194
charity 16, 142, 205, 207–8, 223
chimera 88, 199, 221–2
Christ 47, 80, 97–9, 101–4, 108–9, 172, 214–16, 220
Christianity 48–9, 93, 98, 101, 103, 108, 110, 215–16
climate 3, 150, 185–6
communism 60
conatus 173, 205–7
correlationism 67–8

272 INDEX

cosmos 23, 47, 51–2
creation 1, 16, 21, 45–9, 52, 79, 87,
  90, 132, 135, 139–40, 142–5,
  163–4, 170, 175, 190–4, 197–9,
  203, 210, 218–20
  act of 218–19
Creator 21, 51, 190, 209–10
creatural 9, 164, 167, 217–21, 223
creatures 1, 9–10, 16, 142, 144, 157,
  163, 167, 172–3, 190–3, 198,
  209, 216, 218–20, 223, 225
culture 8, 14, 22, 49, 51, 106, 136,
  185–7, 209
Cunningham, Conor 203

Dasein 177–80, 183
death 2, 14–15, 23, 101, 139–45, 147,
  162, 185–6, 209, 214, 224
decision 64–5, 67, 96, 100, 123, 152
declension type 6, 18, 46–7, 49–50, 53
Deleuze, and Félix Guattari 62, 83–4,
  114, 224
Deleuze, Gilles 3–4, 39, 63, 66, 78,
  82–4, 96, 114, 136, 201, 205–6,
  219
democracy
  real 2, 99
  (of) thought 2, 13, 59–60, 80, 99,
  115, 159
Derrida, Jacques 66, 157, 162
determination 77–8, 88, 114, 159, 178
  -in-the-last-instance 80–1, 86–7, 89
  theological 101, 122, 200
  theological over- 129, 132
dualism 56, 78, 89, 98, 113, 136, 152,
  214
duality 23, 89, 92, 97, 107, 129, 134,
  186, 212
dyad of God and nature 208–9, 211

earth 1, 10, 22–5, 36, 47–51, 104,
  119–21, 128, 132, 135, 143–4,
  147, 175, 183, 185–7, 208, 210,
  213, 216, 222–4

ecological crisis 3, 27, 37, 45, 47–8,
  50, 53–6, 163, 170
ecological degradation 48, 51
ecological imperialism 151
ecological restoration, practice of 21, 151
ecologism 30, 52, 116
ecology
  deep 29–30, 39
  ecosystem 127–8
  history of 22, 53
  human 49, 121
ecosophy 29, 31–2
ecosystem 7–9, 13, 19–21, 24–5, 31,
  36, 43–4, 53–5, 60, 113–17,
  125–32, 134–9, 144–51, 153–5,
  162–4, 169–70, 175–6, 182–4,
  187, 196
  analogical 198
  concept 19–20, 55, 127–30, 182
  individual 25, 135
  philosophical 114–15
  single transcendent 60
  structure 149, 151
energy
  definition of 146, 148
  flows 42, 114, 129–30, 145–7, 150,
  223
energy exchange 19, 31, 114–15, 127,
  137, 145, 147, 170, 176, 182
environment 20–1, 30–2, 37, 47, 116,
  121, 123, 126, 129, 137, 147, 150,
  159–61, 183
environmental thought 37, 39–40, 45,
  47, 152, 168, 176
epochē 34–5
equilibrium 21, 153–5
equivocation 199–200, 207–8, 222
essence 32, 42, 66, 75–6, 79, 84–5,
  104, 108, 162, 180, 194, 200–3,
  207, 210
essence of philosophy 65, 86
ethics 15, 27–9, 39, 45, 47, 56, 79–80,
  83, 110, 136, 140, 199–202, 220,
  225

INDEX

Christian 47
environmental 27–9, 33, 38–9, 90
event 3, 67, 84, 101, 108, 180, 204
experience 4, 39, 64–5, 73, 76, 80–1,
  83, 85, 91, 97, 116, 141, 143, 163,
  171–2, 186, 212, 222–3, 225

fabulation 163, 206, 219–20, 223, 225
fiction 118, 126, 209, 218–19, 222
foundation 72, 120–1, 157, 211–12,
  216, 222
fourfold 8, 168, 183–7, 209, 216, 223

Gaia hypothesis 24, 38, 52, 175–6
generic 56, 81, 87, 91–3, 108–10, 123,
  133, 135, 187
  ecology 110, 122, 136
  ecosystem 184
  secular 62, 105–7
  truth 81
generic matrix 56, 90, 93–4, 107–8,
  116, 145
God 1, 8–9, 15–16, 37–8, 46, 50–1, 53,
  61, 102, 105, 132, 140–4, 162–4,
  172–4, 176, 186–7, 189–94,
  196–205, 207–11, 216–22
  knowledge of 192–4, 202–3
Golley, Frank B. 128, 130
Goodchild, Philip 3, 173–4, 198,
  207
grace 142–3, 145, 203–4, 208
Grant, Iain Hamilton 39–42, 83

Hadot, Pierre 14, 140
Harman, Graham 184
Heidegger, Martin 5, 8, 29, 37–8, 54,
  63, 66, 83, 85, 96, 168, 176–80,
  183–6, 202, 210
heresy 75, 80, 95, 97–8, 101–2, 104,
  106, 108, 215
hierarchy 4, 21, 74, 79, 81–2, 87, 132,
  138, 161
Husserl, Edmund 5, 27, 32–9, 60–1,
  66, 72

identity 7, 13–16, 26, 41, 43, 51, 62,
  87–9, 98, 106–7, 109, 130–1,
  133–4, 136, 146–7, 154–5, 185–7,
  192, 201–3, 209–10
identity of philosophy 62, 179–80
immanence 4–5, 9, 64–6, 77–8,
  84, 86, 91, 104, 107, 109, 114,
  116–17, 121, 125, 127, 129,
  135–6, 139–40, 207–8, 212
  absolute 4, 104, 209–10
  plane of 83
  radical 4–6, 9–10, 14, 65, 76–8,
  81, 86–7, 97–9, 103, 108, 115,
  130, 136, 154, 163, 167, 211–13,
  219, 221–5
in-One 133, 146, 184, 187, 189, 208,
  219
  One- 78, 86
  vision- 45, 65, 70, 76–8, 80, 86–7,
  100, 149
inflection type 6, 18, 49–51, 53–4
Islam, Ismaili 174, 187, 211–12, 216,
  221–2, 224–5

Jambet, Christian 160, 211, 213–16,
  224
Job 140–4, 214
justice 16, 76, 141, 205, 207

Lardreau, Guy 213–16
Laruelle, François 2–4, 6–8, 14, 42, 56,
  59–70, 73–93, 95–105, 107–10,
  113, 115–23, 125, 131, 146, 159,
  163–4, 167, 174, 215, 221–2
Latour, Bruno 6–7, 117, 119, 157–61,
  222
Lévêque, Christian 43, 122, 127, 147,
  150
life
  dialectic of 139–40, 142, 145
  diversity of 135 *see also*
    biodiversity
Llewelyn, John 37–8
Lubac, Henri de 204

Man, radical immanence of 98–9,
        101, 215
material
    ecological 38, 62, 164
    quasi-ecological 38
    religious 62, 97–9, 108
    theological 55, 62, 100, 167
materialism 2, 5–7, 9, 110, 143, 167,
        208, 223
    speculative 67–8
matheme 177, 180–2
matter
    exchange of 145, 149–50, 196, 198
    ultimate 195–6
McFague, Sallie 50–2, 54
Meillassoux, Quentin 39, 64–5,
        67–72, 95
Messiah 108, 143, 215, 220–1, 223
messianic act 211, 214
messianism 91, 93, 209, 211–12, 216
metaphilosophy 59, 62, 73, 83, 118
metaphysics 5, 7, 28, 36, 39, 45, 56,
        68, 70, 73–4, 83, 177, 180, 205,
        218
Milbank, John 106, 219
Morton, Timothy 7, 117, 126, 157,
        160–3, 222
Mullarkey, John 59

Naess, Arne 29–32, 125
naturalism 2, 4, 6–10, 90, 125, 139–40,
        144, 167, 187, 189, 198, 202, 209,
        223–4
nature
    creatural 222–3
    ecological 20, 39
    ecology without 161–2
    perversity of 8–9, 13–15, 25, 41,
        44, 55–6, 61, 144, 167, 197
    philosophy of 18, 26, 28, 39, 41,
        55, 62, 67, 113, 168, 180, 183
    real identity of 183, 210
    theological conceptions of 8, 46,
        135, 164

negation 69, 72, 78, 101–2, 162, 191,
        193, 198, 212, 225
Negri, Antonio 140–4, 214
niche 46, 113, 136–41, 144–5, 153,
        170, 173, 176, 178, 182–4, 219
non-philosophy
    method of 7
    practice of 7, 62, 69, 73–4, 78, 81, 96
    waves of 90, 93
non-religion 102, 105–6
non-theology 7, 62, 72, 99–100,
        102–3, 107–8, 174
non-thetic transcendence 167–8, 208,
        212, 221, 223–4
normativity 9, 135–6
Northcott, Michael 47–8, 50, 52–4
nouvelle théologie 204

object
    of knowledge 85, 122, 213
    real 71
Odum, Eugene and Thomas 128–9, 146
ontology 5, 29, 76, 82–3, 85, 93, 139,
        163, 178–81, 190, 199–200, 225
organicism 19, 83, 128

parallelism 200–2
phenomenology 4, 18, 32–4, 38
philo-fiction 7, 56, 70, 73, 77, 91,
        110, 118, 126
philosophical decision 7, 59, 61–71,
        75–6, 79–80, 86–8, 92, 100,
        107, 113
    theory of 63, 68
philosophy
    Christian 84, 105
    environmental 18, 20, 26–9, 38–9,
        45–6, 55, 59, 66, 151
    first 27, 85, 91
philosophy and ecology, unified
    theory of 115
physis 41, 178–9, 183, 210
piety 3, 9, 99, 108, 170–4, 189, 192, 207
Plato's cave 34, 159–60

INDEX

political ecology 117, 119–20, 122–3, 131, 158–61
Popper, Karl 43–4
protest 10, 139–41, 214, 221
pure nature 202–4, 207–9, 219

radical immanence of nature 9, 217, 219, 223–4
Radkau, Joachim 152
Real 4–5, 14–15, 25, 28, 30, 55–6, 64–81, 84, 86–7, 91, 93, 95–104, 107–8, 110, 116–18, 129, 136, 174, 218–20, 222–5
-One 77–8, 81, 86, 89, 93, 95, 97, 99, 104, 106, 115–16, 134, 163, 210–12, 219
realism 204
rebellion 104–6, 213, 216
regional knowledges 5, 13–14, 51, 55, 59, 81, 89, 93
resilience 60, 115, 117, 135, 151–2, 163
thinking 152, 154–5
resurrection 101, 209–11
revolution 108, 213–14
Ruether, Rosemary Radford 52, 54

Salt, Brian 115, 152–4
Schelling, Friedrich Wilhelm Joseph 5, 40–2
science
autonomy of 84, 86
division of 86, 164
modern 177, 179
philosophy of 33, 44, 82–3, 100, 131
posture of 65, 86
scientism 56, 73, 87, 89–90, 117, 131, 161
secularism 105–8
sékommça 209, 214, 216, 219, 224–5
Sijistânî, Abu Ya'qub al- 8, 164, 168, 175, 209–213, 224
social-ecological systems 152–3
species 9, 13, 22, 43, 123, 126–7, 130, 132–5, 137–9, 145–6, 150, 153, 163–4, 169, 182, 192

dominant 113, 138, 175, 182–3
hybrid 133
Speculative Realism 6, 67
Spinoza, Benedict de 3, 8, 15–17, 29, 66, 87, 164, 168, 173–5, 187, 198–208, 215, 221, 223
substance 40, 136, 194, 200–3
immaterial 194
suffering 2–3, 142–4, 162, 220
sufficient philosophy, principle of 64, 70, 74–5, 100
sufficient theology, principle of 66, 90, 100–2, 189
supernatural 203–5, 221
suspension, realist 70–2
system of energy exchange 19, 147

Tansley, Arthur 19–20, 128–30
Taylor, Paul W. 28–9, 31–2
theology
eco- 45, 54–5
environmental 18, 24, 46–7, 63, 66, 95, 175
Islam 172
liberation 6, 54
natural 47, 218
negative 72, 102, 162, 191, 213
thought
divisions of 163
ecological perspective 148
generic form of 56, 93
image of 20, 23, 149, 164
immanental 90
non-philosophical 88, 101
non-theological 99
Toscano, Alberto 258f107
tradition 54, 102–5, 174
transcendence 5, 47, 65–6, 78, 86, 99, 107, 114, 117, 144, 168, 170, 172, 186, 197–8, 201, 208, 219, 221–2
absolute 143, 221
relative 5

transcendent 5, 9, 21, 65–6, 78, 86,
105, 114, 116, 122, 136, 145, 157,
173, 187, 191, 208
transcendental 35, 40, 61, 65, 75, 77,
84, 86–7, 109, 114, 116, 212
transcendental epochē 35
truth-procedures 180–2
Ṭūsī, Naṣīr al-Dīn al- 8, 164, 187,
224

unified theory 7, 9, 54, 73, 75, 79,
90–1, 99, 107, 110, 116–17, 125,
161, 216–17
univocal predication 192–3

unveiling 184, 186, 190, 198

Velde, Rudi te 190
violence 10, 47, 61, 107, 139–40,
144–5, 197, 221, 223

Whistler, Daniel 233f53
Wilson, Edward O. 42–5, 132–4
world
metaphysical concept of 168, 178
natural 28, 50, 152, 171, 173

Žižek, Slavoj 5, 40–3, 119, 138
Zourabichvili, François 199, 221

CPSIA information can be obtained
at www.ICGtesting.com
Printed in the USA
LVOW13s1447270117

522429LV00004B/6/P